KV-335-247

what if? 2

Also by Randall Munroe

What If?
Thing Explainer
How To

ADDITIONAL SERIOUS SCIENTIFIC ANSWERS

what if? 2

TO ABSURD HYPOTHETICAL QUESTIONS

RANDALL MUNROE

JOHN MURRAY

First published in Great Britain in 2022
by John Murray (Publishers)
An Hachette UK company

1

A CIP catalogue record for this title is available from the British Library

Hardback ISBN 9781473680623
Trade Paperback ISBN 9781473680630
eBook ISBN 9781473680654
Exclusive edition ISBN 9781399804233

Book design by Christina Gleason

Printed and bound in Great Britain by CPI Group (UK) Ltd.

John Murray policy is to use papers that are natural, renewable
and recyclable products and made from wood grown in sustainable forests.
The logging and manufacturing processes are expected to conform to
the environmental regulations of the country of origin.

John Murray (Publishers)
Carmelite House
50 Victoria Embankment
London EC4Y 0DZ

www.johnmurraypress.co.uk

Questions

INTRODUCTION ix

1. Soupiter 1
2. Helicopter Ride 6
3. Dangerously Cold 11
4. Ironic Vaporization 16
5. Cosmic Road Trip 22
6. Pigeon Chair 26

S SHORT ANSWERS #1 30

7. T. Rex Calories 37
8. Geyser 40
9. Pew, Pew, Pew 43
10. Reading Every Book 47

W WEIRD & WORRYING #1 51

11. Banana Church 52
12. Catch! 55
13. Lose Weight the Slow and Incredibly Difficult Way 59
14. Paint the Earth 66
15. Jupiter Comes to Town 70
16. Star Sand 74
17. Swing Set 78
18. Airliner Catapult 83

S SHORT ANSWERS #2 88

19. Slow Dinosaur Apocalypse 98
20. Elemental Worlds 103

21. **One-Second Day** 108
22. **Billion-Story Building** 111
23. **$2 Undecillion Lawsuit** 120
24. **Star Ownership** 124
25. **Tire Rubber** 128
26. **Plastic Dinosaurs** 131
SHORT ANSWERS #3 135
27. **Suction Aquarium** 143
28. **Earth Eye** 149
29. **Build Rome in a Day** 154
30. **Mariana Trench Tube** 159
31. **Expensive Shoebox** 164
32. **MRI Compass** 168
33. **Ancestor Fraction** 172
34. **Bird Car** 176
35. **No-Rules NASCAR** 180
WEIRD & WORRYING #2 185
36. **Vacuum Tube Smartphone** 186
37. **Laser Umbrella** 192
38. **Eat a Cloud** 195
39. **Tall Sunsets** 199
40. **Lava Lamp** 202
41. **Sisyphean Refrigerators** 206
42. **Blood Alcohol** 210
43. **Basketball Earth** 214
44. **Spiders vs. the Sun** 217

45. **Inhale a Person** 220
46. **Candy Crush Lightning** 223
Ⓢ **SHORT ANSWERS #4** 226
47. **Toasty Warm** 234
48. **Proton Earth, Electron Moon** 236
49. **Eyeball** 241
50. **Japan Runs an Errand** 244
51. **Fire from Moonlight** 249
52. **Read All the Laws** 255
Ⓦ **WEIRD & WORRYING #3** 262
53. **Saliva Pool** 263
54. **Snowball** 268
55. **Niagara Straw** 272
56. **Walking Backward in Time** 277
57. **Ammonia Tube** 284
58. **Earth-Moon Fire Pole** 287
Ⓢ **SHORT ANSWERS #5** 298
59. **Global Snow** 305
60. **Dog Overload** 308
61. **Into the Sun** 314
62. **Sunscreen** 318
63. **Walking on the Sun** 323
64. **Lemon Drops and Gumdrops** 329
ACKNOWLEDGMENTS 335
REFERENCES 337
INDEX 349

Disclaimer

Do not try any of this at home.

The author of this book is an internet cartoonist, not a health or safety expert. He likes it when things catch fire or explode, which means he does not have your best interests in mind. The publisher and the author disclaim responsibility for any adverse effects resulting, directly or indirectly, from information contained in this book.

Introduction

I like ridiculous questions because nobody is expected to know the answer, which means it's okay to be confused.

I studied physics in college, so there's a lot of stuff I feel like I'm supposed to know—like the mass of an electron or why your hair sticks up when you rub a balloon against it. If you ask me how much an electron weighs, I feel a little rush of anxiety, like it's a pop quiz and I'm going to be in trouble if I don't know the answer without looking it up.

But if you ask me how much all the electrons in a bottlenose dolphin weigh, that's a different situation. No one knows that number off the top of their head—unless they have an *extremely* cool job—which means it's okay to feel confused and a little silly and take some time to look stuff up. (The answer, in case anyone ever asks you, is about half a pound.)

Sometimes simple questions turn out to be unexpectedly hard. Why *does* your hair stand on end when you rub a balloon on it, anyway? The usual answer from science class is that electrons are transferred from your hair to the balloon, leaving your hair positively charged. The charged hairs repel each other and stick out.

Except . . . why do electrons get transferred from the hair to the balloon? Why don't they go the other way?

That's a great question, and the answer is that no one knows. Physicists don't have a good general theory for why some materials shed electrons from their surfaces on contact while other materials pick them up. This phenomenon, called triboelectric charging, is an area of cutting-edge research.

The same kind of science is used to answer serious questions and silly ones. Triboelectric charging is important to understanding how lightning forms in storms. Counting the number of subatomic particles in an organism is something physicists do when modeling radiation hazards. Trying to answer silly questions can take you through some serious science.

And even if the answers aren't useful for anything, knowing them is fun. The book you're holding weighs about as much as the electrons in two dolphins. That information probably isn't useful for anything, but I hope you enjoy it, anyway.

what if? 2

1. SOUPITER

What would happen if the Solar System was filled with soup out to Jupiter?

—Amelia, age 5

Please make sure everyone is safely out of the Solar System before you fill it with soup.

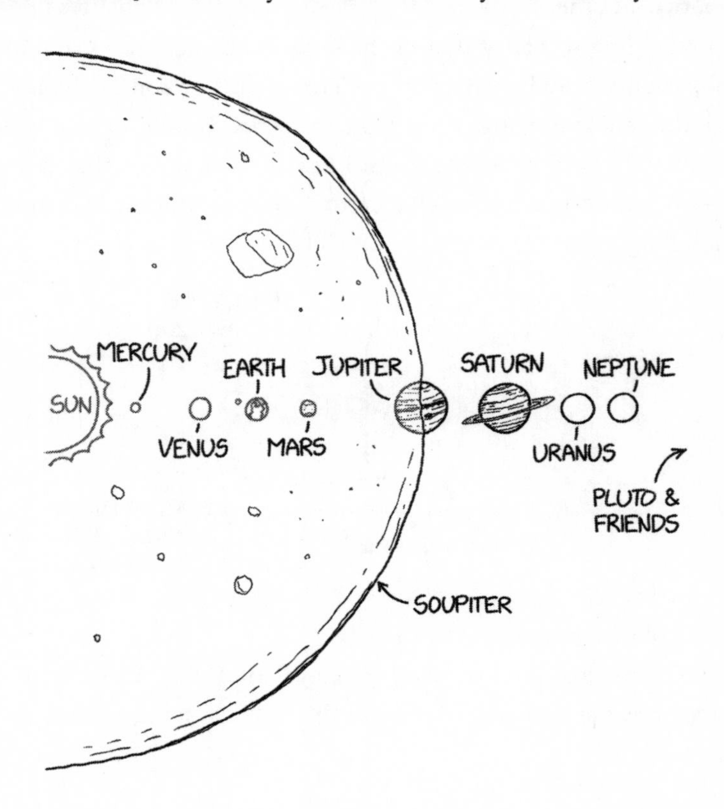

If the Solar System were full of soup out to Jupiter, things might be okay for some people for a few minutes. Then, for the next half hour, things would definitely not be okay for anyone. After that, time would end.

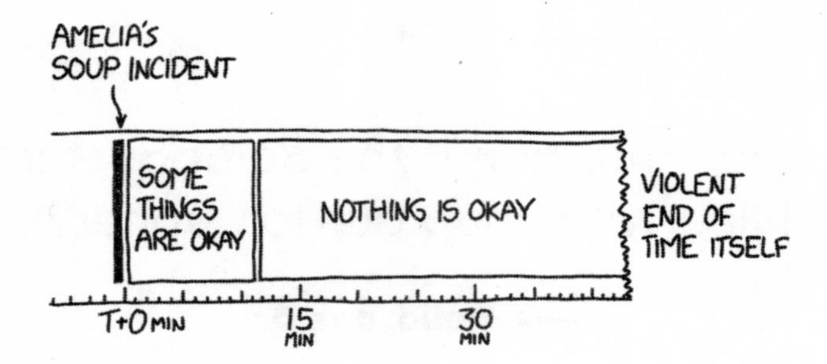

Filling the Solar System would take about 2×10^{39} liters of soup. If the soup is tomato, that works out to about 10^{42} calories worth, more energy than the Sun has put out over its entire lifetime.

The soup would be so heavy that nothing would be able to escape its enormous gravitational pull; it would be a black hole. The event horizon of the black hole, the region where the pull is too strong for light to escape, would extend to the orbit of Uranus. Pluto would be outside the event horizon at first, but that doesn't mean it would escape. It would just have a chance to broadcast out a radio message before being vacuumed up.

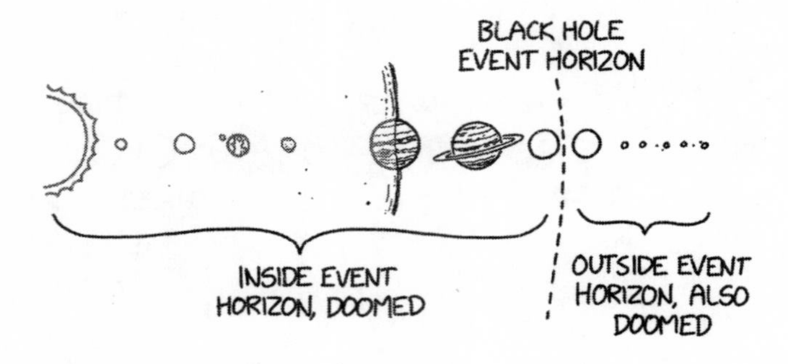

What would the soup look like from inside?

You wouldn't want to stand on the surface of the Earth. Even if we assume the soup is rotating in sync with the planets in the Solar System, with little whirlpools surrounding each planet so the soup is stationary where it touches their surfaces, the pressure due to the Earth's gravity would crush anyone on the planet within seconds. Earth's gravity may not be as strong as a black hole's, but it's more than enough to pull an ocean of soup down hard enough to squish you. After all, the pressure of our regular water oceans under Earth's gravity can do that, and Amelia's soup is a lot deeper than the ocean.

If you were floating between the planets, away from Earth's gravity, you'd actually be okay for a little while, which is kind of weird. Even if the soup didn't kill you, you'd still be inside a black hole. Shouldn't you die instantly from . . . something?

Strangely enough, no! Normally, when you get close to a black hole, tidal forces tear you apart. But tidal forces are weaker for larger black holes, and the Jupiter Soup black hole would be about 1/500th the mass of the Milky Way. That's a monster even by astronomical standards—it would be comparable in size to the largest known black holes. Amelia's souper-massive black hole would be large enough that the different parts of your body would experience about the same pull, so you wouldn't be able to feel any tidal forces.

Even though you wouldn't be able to *feel* the soup's gravitational pull, it would still accelerate you, and you would immediately begin to plunge toward the center. After a second had passed, you'd have fallen 20 kilometers and you'd be traveling at 40 kilometers per second, faster than most spacecraft. But since the soup would be falling along with you, you'd feel like nothing was wrong.

I FEEL PRETTY NORMAL!
NO PROBLEMS HERE!
WE'RE FLOATING IN SOUP AND I THINK THE SUN JUST COLLAPSED.

As the soup collapsed inward toward the center of the Solar System, its molecules would be squeezed closer together and the pressure would rise. It would take a few minutes for this pressure to build up to levels that would crush you. If you were in some kind of a soup bathyscaphe, the pressure vessels that people use to visit deep ocean trenches, you could conceivably last for 10 or 15 minutes.

There would be nothing you could do to escape the soup. Everything inside it would flow inward toward the singularity. In the regular universe, we're all dragged forward through time with no way to stop or back up. Inside a black hole's event horizon, in a sense time stops flowing *forward* and starts flowing *inward*. All time lines converge toward the center.

From the point of view of an unlucky observer inside our black hole, it would take about half an hour for the soup and everything in it to fall to the center. After that, our definition of time—and our understanding of physics in general—breaks down.

Outside the soup, time would continue passing and problems would keep happening. The black hole of soup would start slurping up the rest of the Solar System, starting with Pluto almost immediately, and the Kuiper belt shortly thereafter. Over the course of the next few thousand years, the black hole would cut a large swath through the Milky Way, gobbling up stars and scattering more in all directions.

This leaves us with one more question: What kind of soup is this, anyway?

If Amelia fills the Solar System with broth, and there are planets floating in it, is it planet soup? If there are already noodles in the soup, does it become planet-and-noodle soup, or are the planets more like croutons? If you make a noodle soup, then someone sprinkles some rocks and dirt in it, is it really noodle-and-dirt soup, or is it just noodle soup that got dirty? Does the presence of the Sun make this star soup?

The internet loves arguing about soup categorization. Luckily, physics can settle the debate in this particular case. It's believed that black holes don't retain the characteristics of the matter that goes into them. Physicists call this the *no hair theorem*, because it says that black holes don't have any distinguishing traits or defining characteristics. Other than a handful of simple variables like mass, spin, and electric charge, all black holes are identical.

In other words, it doesn't matter what kind of ingredients you put into a black hole soup. The recipe always turns out the same in the end.

2. HELICOPTER RIDE

What if you were hanging on a helicopter blade by your hands and then someone turned it on?

—Corban Blanset

You may be picturing a cool movie action scene like this:

If so, you're going to be disappointed, because what would actually happen would be more like this:

Helicopter rotors take a little while to get up to speed. Once the rotor starts moving, it might take 10 or 15 seconds for it to make its first full turn, so you'd have an uncomfortable amount of time to make eye contact with the pilot before you rotated out of view.

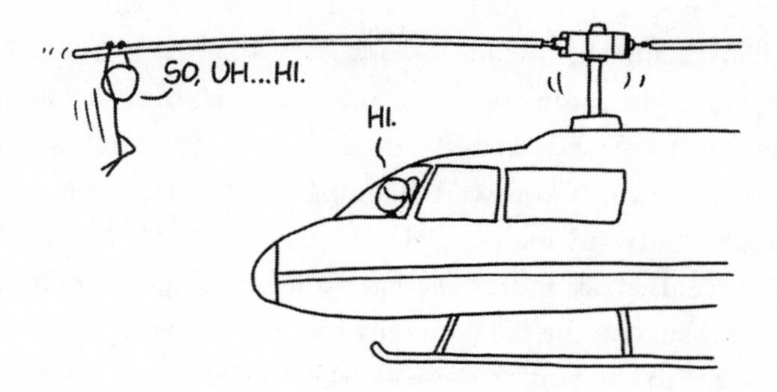

Luckily, you probably won't have to pass in front of the pilot a second time, because you'll fall off embarrassingly quickly.

Hanging on to the smooth surface of the blade would be hard enough when it was standing still, but even if you found a comfortable handhold, you'd probably lose your grip before the blade finished a single turn.

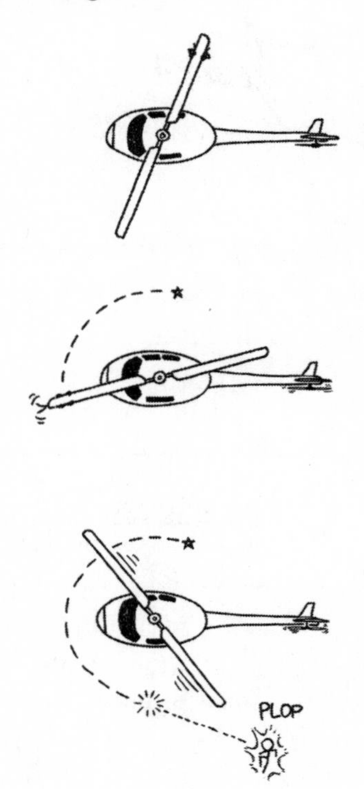

Helicopter blades are pretty big, which makes them look like they're moving more slowly than they really are. We're not used to large objects moving around that fast. When a helicopter is sitting on the pad with the rotor revolving slowly, it may look pretty gentle, like a dangling mobile rotating over a baby's crib. But if you tried to hang on to the end of the rotor, you'd find yourself flung outward surprisingly hard.

It might take 5 to 10 seconds from the time when the rotor starts moving to when it makes its first half-turn. If you were hanging on, by that point you'd already be swinging noticeably outward and you'd feel an extra 10 or 20 pounds of weight from the centrifugal force. Luckily, most helicopter rotors are close enough to the ground that you'd probably survive the fall with only minor injuries and bruised dignity.

If you do manage to hang on, things will get worse very fast. By the time the blade makes one full turn,* the centrifugal force will be pulling on you even harder than gravity, causing you to swing way outward. The extra force would be the equivalent of the weight of another person clinging to you.

Even if you had a really good grip, you'd probably struggle to hang on. If you wanted to ride the rotor all the way around, you'd need to arrange some kind of system to keep your hands attached to the blade.

* Definitely pick a helicopter that has a sufficient gap between the tail rotor and the main blade, or you'll need to get really good at doing pull-ups at the right time.

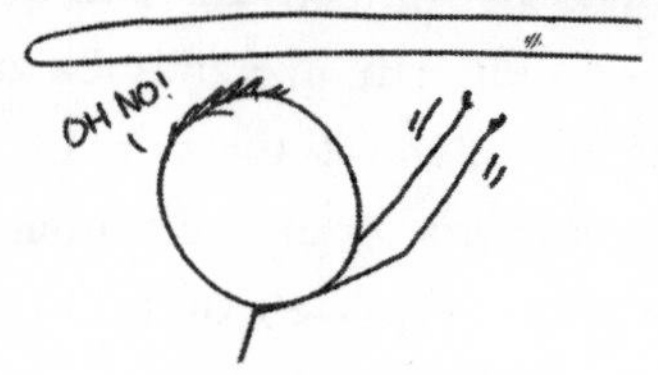

If the rotor kept accelerating at its normal rate, and you somehow stayed attached, then after another full rotation you'd be swinging almost straight outward, with your hands trying to support many times your own body weight. If you hung on for 20 seconds, the rotor would be making one revolution per second, putting several tons of force on your hands. After 30 seconds, you'd have lost your grip on the helicopter one way or another. If your hands stay attached to the rotor, they won't stay attached to your body.

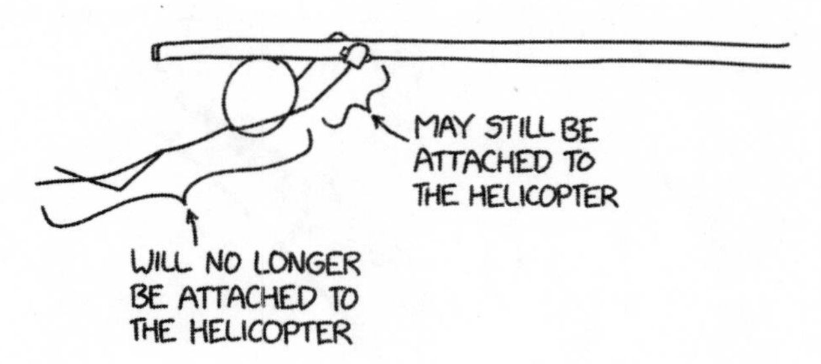

This experience won't be any more pleasant for the helicopter than for you. The rotor wouldn't be able to keep accelerating like it would during a normal start-up.

After all, if your hands are experiencing this much force, then so is the helicopter. A helicopter blade is designed to handle many tons of tension, but that tension is carefully balanced between the blades. If one blade is exerting more force than the other, it will yank the helicopter back and forth, like an unbalanced washing machine.

Adding just a few ounces of weight to the base of a blade can cause (or cancel out) uncomfortably strong vibrations. Adding a human-size weight to the end of a blade would cause the helicopter to flip itself over and tear itself apart long before it got up to speed.

Come to think of it, maybe this *would* make a good movie action scene. You know the scene where the villain's helicopter is escaping, and the hero runs and jumps and dangles from the landing skids?

If the hero really wants to keep the villain from escaping . . .

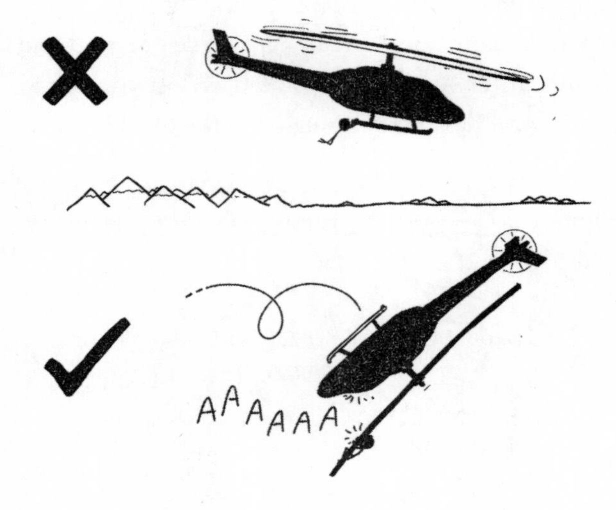

. . . they should just grab a little higher.

3. DANGEROUSLY COLD

Would there be any danger from standing next to a large object that was 0 Kelvin?

—Christopher

So you've decided to install an ultracold cube of iron in your living room.

First of all, definitely don't touch it. As long as you resist the urge to touch it, you probably won't suffer any immediate harm.

Cold things and hot things are different.[citation needed] Standing near a hot object can kill you very fast—for more on this, flip to basically any other random page of this book—but standing near a cold thing won't freeze you instantly. Hot objects emit thermal radiation that heats up things around them, but cold objects don't emit cold radiation. They just sit there.

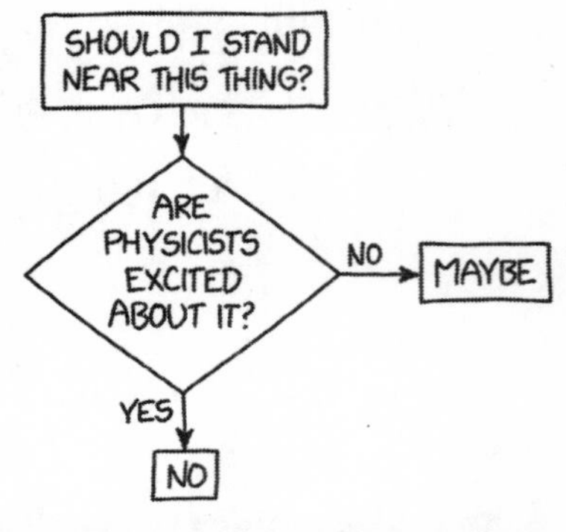

Even though it doesn't give off cold radiation, the *lack* of heat radiation can make you feel cold. Your body, like all warm objects, is constantly radiating heat. Luckily for you, everything around you—like furniture and walls and trees—is *also* radiating heat, and that incoming radiation partly balances out the heat you're losing. We usually measure room temperatures in Fahrenheit or Celsius, but setting our thermostats to Kelvin would make it clearer that most of the stuff in the room has roughly the same absolute heat level—since it's all 250 or 300 degrees Kelvin—so it all radiates heat.

When you stand near something much colder than room temperature, the heat you're losing in that direction isn't balanced by any incoming heat, so that side of your body gets cold much faster. From your point of view, it feels like the object is radiating cold.

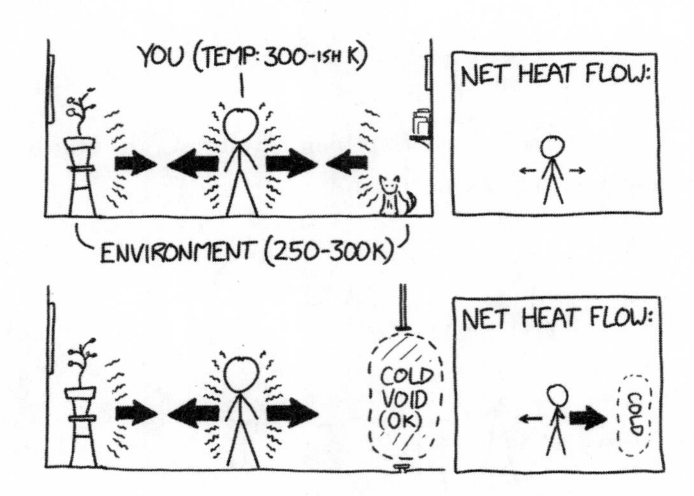

You can feel this "cold radiation" by looking up at the stars on a summer night. Your face will feel cold since your body heat is pouring away into space. If you hold up an umbrella to block your view of the sky, you'll feel warmer—almost as if the umbrella is "blocking the cold" from the sky. This "cold sky" effect can cool things down to below the ambient air temperature. If you leave out a tray of water under a clear sky, it can turn to ice overnight even if the air temperature stays well above freezing.

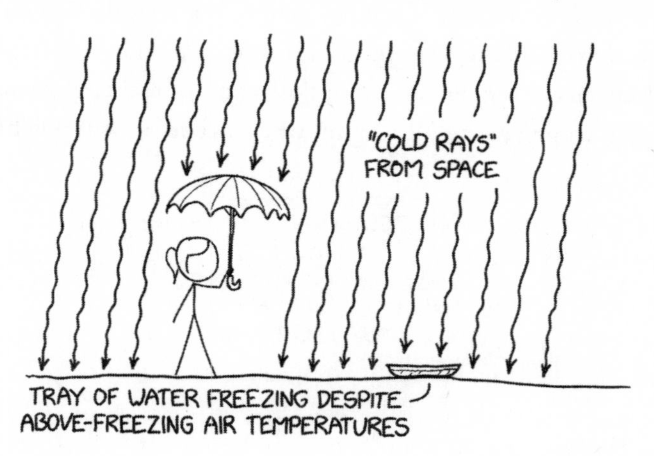

You'll feel chilly standing next to your cube, but not *that* chilly—nothing a good winter coat can't solve. But before you rush to get a cryogenic cube, we need to talk about the air.

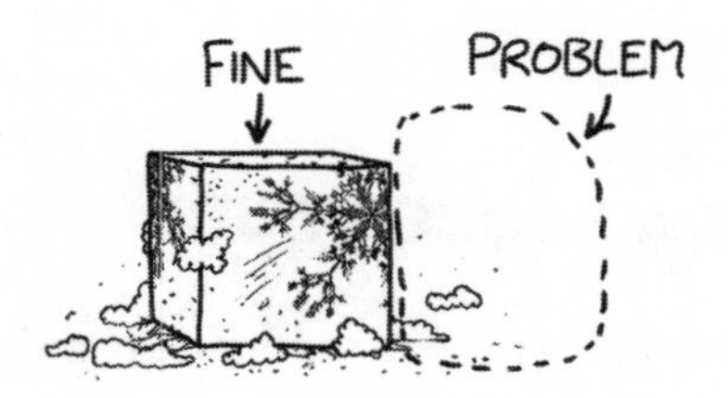

Cold objects can condense the air itself, causing liquid oxygen to collect on their surfaces like dew. If they're cold enough, they can even freeze it solid. Engineers working with cold industrial equipment have to watch out for this oxygen buildup, since liquid oxygen is pretty dangerous stuff. It's highly reactive and tends to cause flammable things to spontaneously ignite. A really cold object can set your house on fire.

One of the biggest hazards of ultracold materials is that they often don't want to *stay* ultracold. When liquid nitrogen or dry ice warm up and turn to gas, they expand a *lot*, often pushing all the regular air out of the room. A bucket of liquid nitrogen can turn into enough nitrogen gas to fill a room, which is bad news if you breathe oxygen.

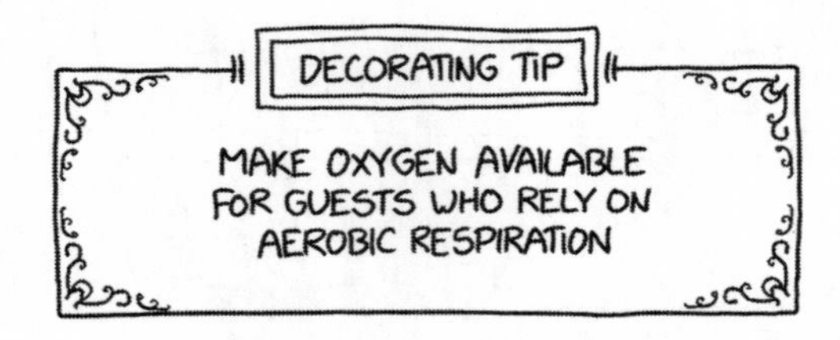

Luckily, iron is solid at room temperature, so you don't have to worry about your cube of iron evaporating. As long as you avoid touching it, keep any oxygen on the surface from coming into contact with anything flammable, and wear a winter coat, you'll probably be fine.

SO YOU'VE DECIDED YOU DON'T WANT A FROZEN CUBE

The cube will take an awfully long time to warm up. It will sit there at cryogenic temperatures for days, soaking up heat from the room while remaining cold enough to freeze the air. Even if you open the windows and run the furnace at full blast to keep the surrounding air as warm as possible, it will take at least a week for the cube to get close to room temperature.

You could try to speed the process up by surrounding the cube with a dozen space heaters—with the help of an electrician, because otherwise you'll blow all the fuses in your house—but it would still take days to warm it up.

If you wanted to thaw out the cube more quickly, you could try pouring water on it. The water would instantly turn to ice, which you could chip away and discard, leaving some of the water's heat behind in the iron. It might take a few bathtubs full of water, but you could use this technique to get the cube up to a reasonable temperature more quickly.

Once the iron reaches room temperature, it will become just another object in your house. Hopefully, you like it where it is—if not, given how hard it would be to move a smooth eight-ton cube, it might be easier for *you* to move instead.

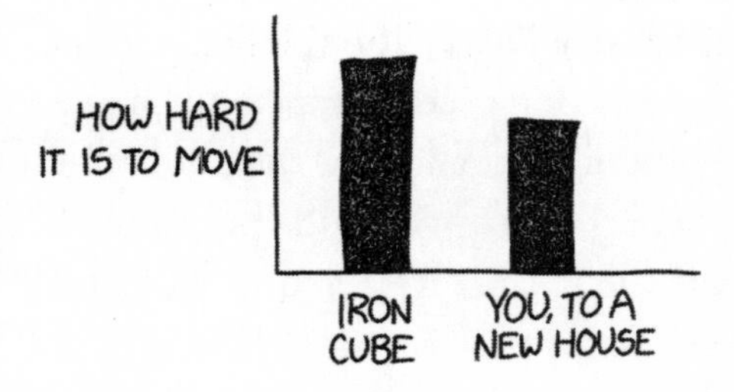

If you don't want to move, and you're looking for another way to get rid of a cube of iron, you could always try adding *more* heat to it.

To find out what happens if you do that, turn to the next chapter.

4. IRONIC VAPORIZATION

What if we somehow evaporated a solid block of iron on earth?

—Cooper C.

So you've decided to evaporate a one-meter cube of iron in your backyard.

Iron can boil and evaporate like anything else, but since its boiling point is so high—roughly 3,000°C—you don't see it happen much in everyday life.

To boil water, you put it in a pot and heat the pot until the water reaches 100°C. Boiling iron is trickier, because what would the pot be made of? Most metals have a melting point below iron's boiling point, so you wouldn't be able to use them to hold boiling iron—they'd melt before the iron started to simmer.

There are a few substances that remain solid slightly above iron's boiling point, like tungsten, tantalum, or carbon, but using them to hold boiling iron is tricky. Getting the iron to boil while keeping the container below its melting point is difficult in practice, and there are chemical problems as well. Iron is chemically troublesome—once it's molten, it tends to react with its container and form alloys.

In real life, when people want to vaporize iron,* they generally don't just put it over a heat source. They either use induction heating to heat the iron with electromagnetic fields or electron beams to vaporize it a little at a time. One nice thing about electron beams is that you can use a magnetic field to curve the beam around, so the really exciting and dangerous stuff happens on the other side of the iron from your delicate equipment.

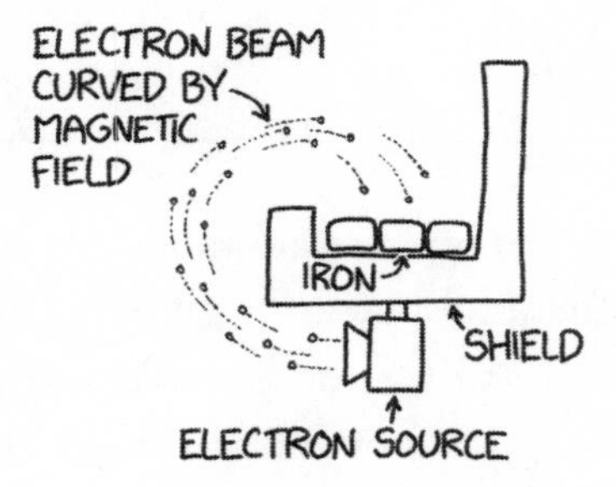

You should be sure to stand on the "shield" side of the apparatus, since the iron evaporation side will have lots of high-energy particles flying away from it. "Stand on the other side from where the physics is happening" is actually a good general rule for scientific equipment.

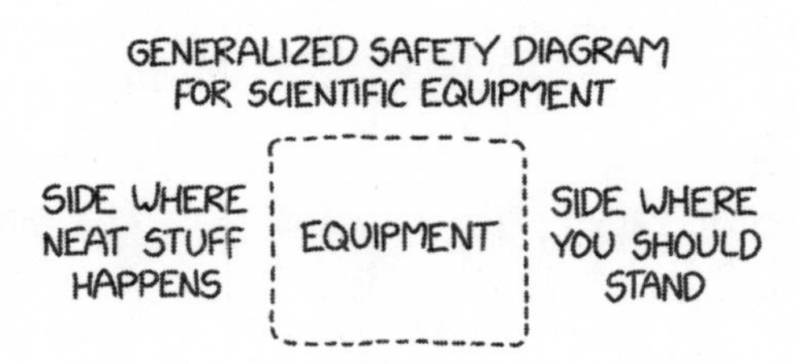

* Usually to use the vapor for metal plating, but maybe sometimes just out of spite.

Once you've built your apparatus to vaporize iron, you'll want to stand back, since vaporizing a 1-meter cube of iron will take about 60 gigajoules of energy. If you vaporize the iron over the course of three hours, your apparatus will have roughly the same total heat output as a raging house fire.*

But your question wasn't whether we could *do* it. It was what the consequences would be, and the answer to that is pretty simple: Your house and yard would catch fire. Then the fire department would show up, and lots of people would be mad at you.

The consequences to the atmosphere are more interesting. You'd be releasing a plume of 8 tons of iron into the atmosphere—what would that do to your surroundings?

It wouldn't have a big effect on the atmosphere as a whole. There's already a lot of iron in the air, most of it in the form of wind-blown dust. Human activities, mostly the burning of fossil fuels, also pump a lot of iron into the air. Based on estimates from a 2009 study by Natalie Mahowald et al., over the three hours it takes to vaporize your 8-ton iron cube, desert winds will blow 30,000 tons of iron into the air, and industrial facilities will add another 1,000 tons.

IRON ADDED TO THE ATMOSPHERE DURING YOUR PROJECT

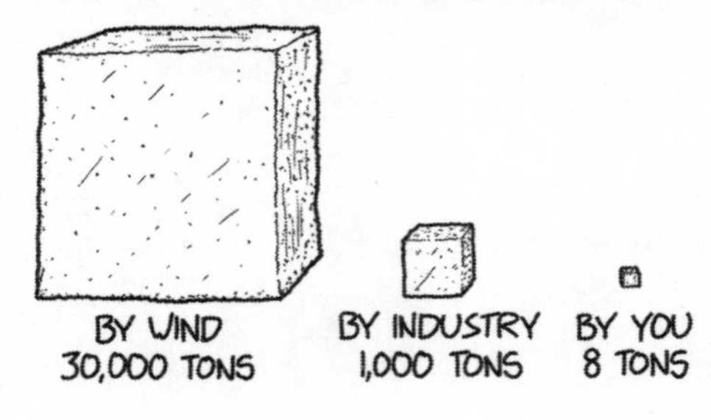

* If you do this project near your actual house, you may find it produces the heat of *two* house fires.

Eight tons of iron might not affect the Earth as a whole, but what about your neighbors? What would the people downwind of you notice, besides the fire trucks? Would they wake up to find everything metal-plated?

To answer these questions, I reached out to Dr. Mahowald, the lead author of the 2009 study and an expert on atmospheric transport of metals.

Dr. Mahowald explained that when you release a plume of iron vapor, the iron rapidly reacts with oxygen in the air to condense into iron oxide particles. "Iron oxide particles aren't particularly hazardous for air quality," she said, although if there are enough of them, they could certainly be bad for your lungs. That's not necessarily because of any properties specific to iron oxide—it's just that your lungs are designed to breathe air.

Eventually, the iron oxide particles would settle out of the air somewhere downwind of your houses, but they wouldn't necessarily cause any serious problems. "They probably wouldn't kill anything," Dr. Mahowald said. "On land, there's a good bit of iron already." But if there was enough of it, she added, it could cover up the vegetation, like the layers of ash downwind of a volcanic eruption. Your neighbors might be annoyed because they have to brush off their car.

Dr. Mahowald said the vaporized iron would contribute to climate change by absorbing small amounts of sunlight and radiating it as heat. But iron in the atmosphere could also help slow down climate change, by fertilizing the ocean and encouraging the growth of algae that pull CO_2 out of the atmosphere. In 1988, oceanographer John Martin famously claimed—in his best supervillain voice—"Give me half a tanker of iron, and I will give you an ice age."

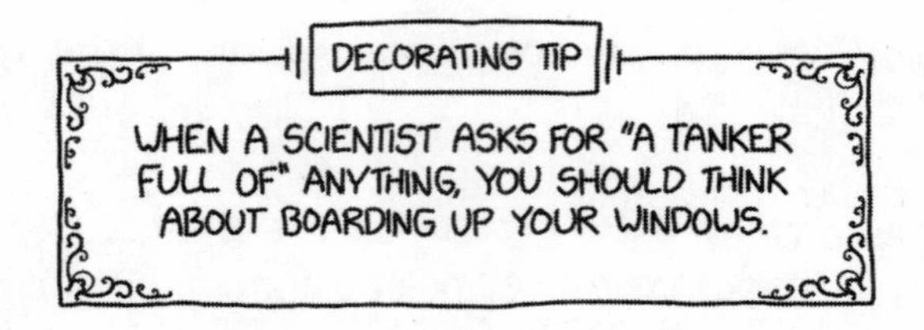

Dr. Martin never became a supervillain[citation needed] and never attempted this plan, but it's doubtful it would have worked. Further research has shown that dumping iron in the ocean is probably not an efficient way to pull carbon out of the air, which is kind of disappointing for supervillains who want to cause an ice age *and* for superheroes who want to stop global warming.

But if you *do* have a block of iron and the means to vaporize it, and you really hate your house, yard, and the gardens of the neighbors who live downwind of you, then I have some great news about your plan.

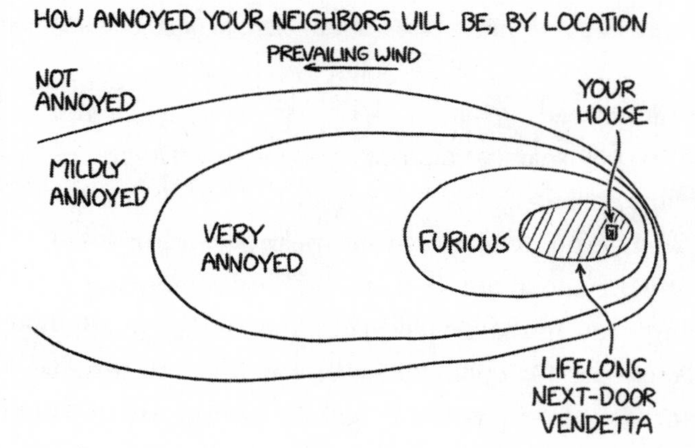

5. COSMIC ROAD TRIP

If the universe stopped expanding right now, how long would it take for a human to drive a car all the way to the edge of the universe?

—Sam H-H

The edge of the observable universe is about 270,000,000,000,000,000,000,000 miles away.

If you drive at a steady 65 miles per hour, it will take you 480,000,000,000,000,000 years—that's 4.8×10^{17}—to get there, or 35 million times the current age of the universe.

This will be a dangerous road trip. I don't mean because of space stuff—we're not worrying about all that—but because driving itself is pretty dangerous. In the United States, the average middle-aged driver suffers about one fatal crash per 100 million miles driven. If someone built a highway out of the Solar System, most drivers wouldn't make it past the asteroid belt. Truck drivers, who are used to driving long distances on highways, have a lower per-mile crash rate than ordinary drivers, but they would still be unlikely to reach Jupiter.

Based on US crash rates, the odds of a driver traveling 46 billion light-years without a crash would be about 1 in $10^{10^{15}}$. That's roughly the same as the probability of a monkey with a typewriter typing out the entire Library of Congress, with no typos, *fifty times in a row.* You'll want a self-driving car, or at least one with one of those alarms that warns you if you drift out of your lane.

The trip would take a lot of fuel. At 33 miles per gallon, it would take a Moon-size sphere of gasoline to reach the edge of the universe.* You'd run through about 30 quintillion oil changes, requiring a container of engine oil the volume of the Arctic Ocean.†

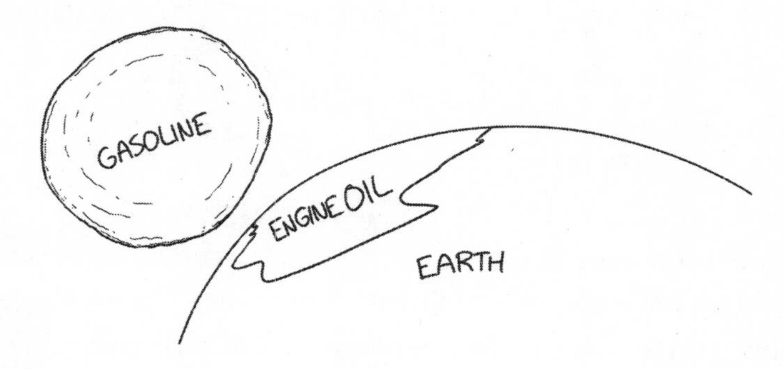

You'd also need 10^{17} tons of snacks. Hopefully, there are a lot of intergalactic rest areas, or your trunk is going to be pretty full.

* As of 2021, NASA's *New Horizons* spacecraft has traveled about 5 billion miles on a budget of about $850 million, which works out to 17 cents per mile—pretty similar to the cost of gas and snacks on a road trip.

† An old piece of advice says that you need to change your oil every 3,000 miles, but most car experts agree that's a myth—modern gasoline engines can comfortably go two or three times that distance between changes.

It's going to be a very long drive, and the scenery won't change much at all. Most of the visible stars will burn out before you even exit the Milky Way galaxy. If you want to try touching a room-temperature star—see chapter 63 for what that would be like—I suggest planning a route that takes you past Kepler-1606. It's 2,800 light-years away, so when you drive past it after 30 billion years, it will have cooled to a comfortable room temperature. It has a planet right now, although it will have probably devoured it by the time you get there.

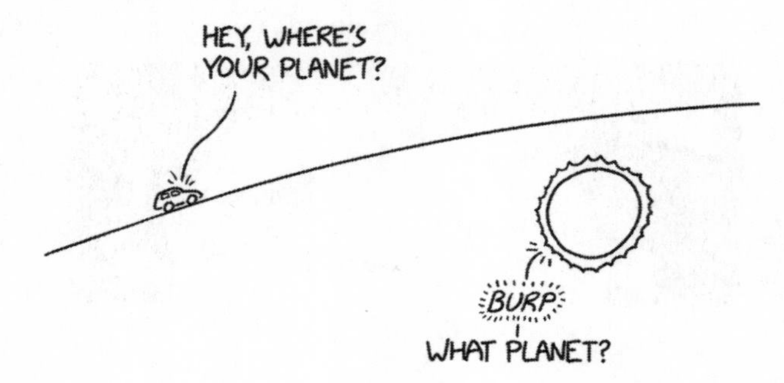

Once the stars have burned out, you'll have to find a new source of entertainment. Even if you bring every audiobook ever recorded and every episode of every podcast, that won't even last you to the edge of the Solar System.

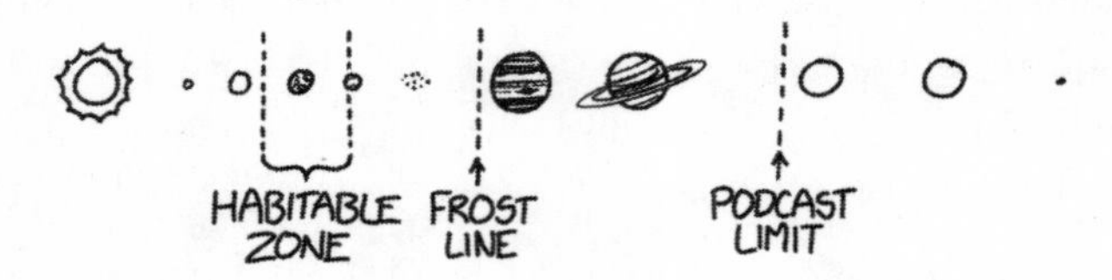

Robin Dunbar famously suggested that the average human maintains about 150 social relationships. The total number of humans who have ever lived is somewhere north of 100 billion. A 10^{17}-year road trip would be long enough to replay the lives of every one of those people in real time—in a sort of unedited documentary—and then *rewatch* every one of those documentaries 150 times, each time with a different commentary track by the 150 people who knew the subject best.

By the time you finished watching this complete documentary of human perspective, you'd still be less than 1 percent of the way to the edge of the universe, so you'd have plenty of time to rewatch the whole project—each human life with all 150 commentary tracks—100 times before you finally arrived.

Once you reached the edge of the observable universe, you could spend another 4.8×10^{17} years driving back home, but since there won't be any Earth to return to—all that will be left are black holes and frozen husks of stars—you might as well keep going.

As far as we know, the edge of the observable universe isn't the edge of the actual universe. It's just the farthest that we're able to see, because there hasn't been time for light to reach us from any farther parts of space. There's no reason to think space itself ends at that particular point, but we don't know how much farther it goes. It might just continue forever. The edge of the observable universe isn't the edge of space, but it's the edge of the map. There's no way to be sure what you'll find when you cross it.

Be sure to pack extra snacks.

6. PIGEON CHAIR

How many pigeons would it require in order to lift the average person and a launch chair to the height of Australia's Q1 skyscraper?

—Nick Evans

Believe it or not, science can answer this question.

In a 2013 study, researchers at the Nanjing University of Aeronautics and Astronautics led by Ting Ting Liu trained pigeons to fly up to a perch while wearing a weighted harness. They found that the average pigeon in their study could take off and fly upward while carrying 124 grams, about 25 percent of its body weight.

The researchers determined that the pigeons could fly better if the weights were slung below their bodies, rather than on their backs, so you would probably want pigeons to lift your chair from above rather than support it from below.

Let's suppose your chair and harnesses weigh 5 kilograms and you weigh 65 kilograms. If you used the pigeons from the 2013 study, it would take a flock of about 600 of them to lift your chair and fly upward with it.

Unfortunately, flying with a load is a lot of work. The pigeons in the 2013 study were able to carry a load 1.4 meters upward to a perch, but they probably wouldn't have been able to fly too much higher than that. Even unencumbered pigeons can only maintain strenuous vertical flight for a few seconds. One 1965 study measured a climb rate of 2.5 m/s for unencumbered pigeons,* so even if we're being optimistic, it seems unlikely that pigeons could lift your chair more than 5 meters.†

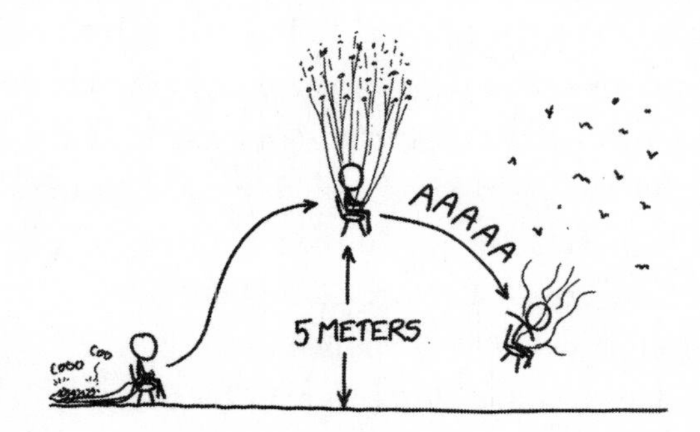

* Here's how the authors of the 1965 study, C. J. Pennycuick and G. A. Parker, describe their method for measuring pigeon vertical flight speed: "Tame pigeons were fed by hand in the open on the flat roof of the laboratory in a corner of the 107 cm. high wall which surrounds the roof. A cine camera was set up level with the top of the wall, pointing into the corner. On the camera being started, a helper rushed at the pigeons, which were thus forced to make a near-vertical climb in order to get over the wall." I love methods sections.

† According to a 2010 study by Angela M. Berg et al., about 25 percent of the pigeon's takeoff acceleration comes from pushing off with its legs. Since they'd be kicking down against the craft to take off, they'd have a lot more work to do with their wings, making these estimates even more optimistic than they already were.

No problem, you might think. If 600 pigeons can lift you the first 5 meters, then you just need to bring another 600 along with you, like the second stage of a rocket, to carry you the next 5 meters when the first flock gets tired. You can bring another 600 for the 5 meters after that and so on. The Q1 is 322 meters high, so about 40,000 pigeons should be able to get you to the top, right?

No. There's a problem with this idea.

Since a pigeon can carry only a quarter of its body weight, it takes four flying pigeons to carry one resting pigeon. That means each "stage" will need at least four times as many pigeons as the one above it. Lifting one person may only take 600 pigeons, but lifting one person *and* 600 resting pigeons would take another 3,000 pigeons.

This exponential growth means that a 9-stage vehicle, able to lift you 45 meters, would need almost 300 million pigeons, roughly equal to the entire global population. Reaching the halfway point would require 1.6×10^{25} pigeons, which would weigh about 8×10^{24} kilograms—more than the Earth itself. At that point, the pigeons wouldn't be pulled down by the Earth's gravity—the Earth would be pulled up by the pigeons' gravity.

The full 65-stage craft to reach the top of the Q1 would weigh 3.5×10^{46} kilograms. That's not just more pigeons than there are on Earth, it's more mass than there is in the galaxy.

A better approach might be to avoid carrying the pigeons with you. After all, pigeons can get up to the top of the skyscraper themselves, so you might as well send them ahead to wait for you there instead of having their friends carry them up with you. If you could train them well enough, you could have them glide along at the appropriate height, then grab you and tug you upward for a few seconds when you reach their altitude. Keep in mind that pigeons can't grab and carry things with their feet, so they'd need little harnesses with aircraft-carrier-style hooks to intercept you.

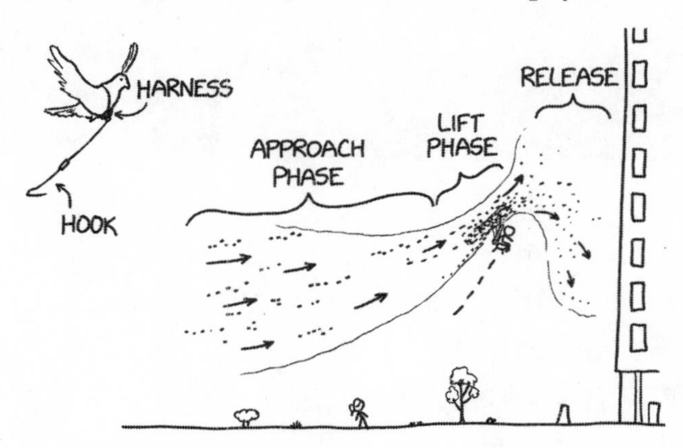

With this arrangement, it's possible you could fly yourself to the top of the tower with just a few tens of thousands of well-trained pigeons. You should probably make sure you have some kind of safety system that will keep you from plunging to your demise every time a falcon flies by and spooks the pigeons.

The craft wouldn't just be more dangerous than an elevator, it would also be a lot harder to pick your destination. You might *plan* to go to the top of the Q1, but once you take off . . .

. . . you'll be completely under the control of anyone with a bag of seeds.

short answers #1

Q What if your blood became liquid uranium? Would you die from radiation, lack of oxygen, or something else?

—Thomas Chattaway

YOU WOULD DIE FROM WHAT WE IN THE MEDICAL PROFESSION CALL NOT-HAVING-ANY-BLOOD-AND-BEING-FULL-OF-MOLTEN-URANIUM SYNDROME.

OR "JEFF'S DISEASE" FOR SHORT.

MAN, POOR JEFF.

Q Could someone have an anime-style attack where they created a sword out of air? I'm not talking about an air blade, but something like cooling the air enough so that you had solid air to attack people.

—Emma from Manhattan

Sure. It would take a whole room full of air, but you could do it.

Studies of solid oxygen suggest that it has mechanical properties similar to soft plastic, growing a bit harder as it gets colder. So if you make your sword out of oxygen, it wouldn't be very strong, it would be hard to sharpen, and it would quickly give your hand frost damage. Nitrogen, which has a slightly higher melting point, wouldn't be much better. But you could do it.

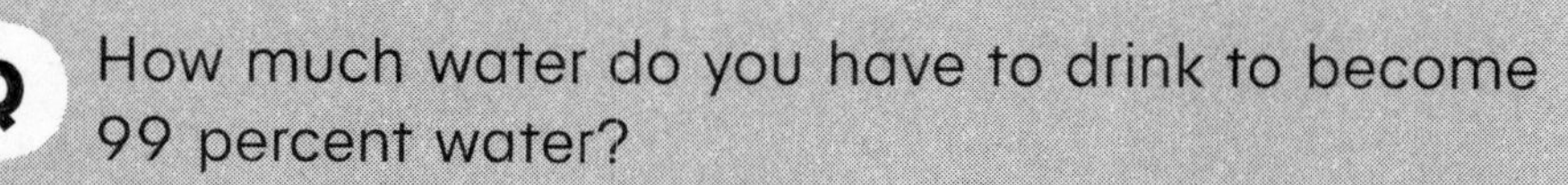

Q How much water do you have to drink to become 99 percent water?

—LyraxH

$$\frac{\text{NEW WATER} + \text{BODY WATER}}{\text{BODY NON-WATER}} = \frac{99}{1}$$

$$\begin{aligned}\text{NEW WATER} &= \frac{99}{1} \times (\text{NON-WATER}) - \text{WATER} \\ &= \frac{99}{1} \times \left(1 - \frac{70}{100}\right) \times 65L - \frac{70}{100} \times 65L \\ &= 29 \times 65L \\ &\approx 1{,}900\,L \approx 500 \text{ GAL}\end{aligned}$$

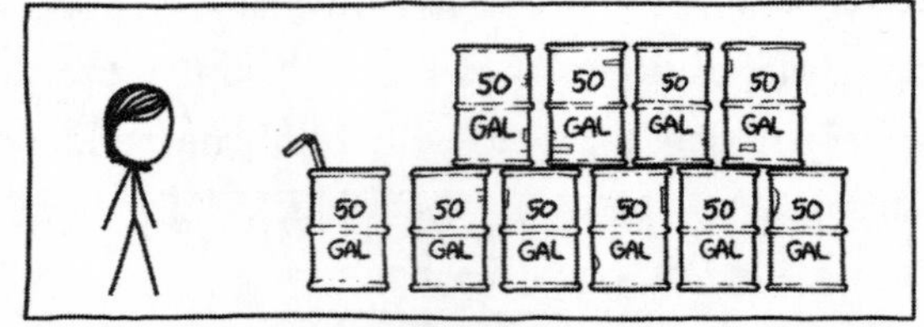

HERE YOU GO!

Q What would we see if we attached a lightweight camera to a balloon and let it fly away?

—Raymond Peng

Q How many calories does Mario burn a day?

—daniel and xavier hovley

MUSHROOMS IN *SUPER MARIO BROS*: 56
CALORIES IN ONE MEDIUM MUSHROOM: 5
TOTAL AVAILABLE CALORIES: 280
SUPER MARIO BROS RELEASE DATE: SEPT 13, 1985
RELEASE OF NEXT MARIO GAME WITH MUSHROOMS: JUNE 3, 1986
INTERVAL: 263 DAYS
CALORIES PER DAY: 1.1

CONCLUSION

MARIO STARVED TO DEATH IN LATE 1985.

Q If a snake unhinged its jaw and swallowed a balloon whole, could/would the balloon carry the snake up?

—Freezachu

> **Q** If you were to jump out of an airplane that was traveling at Mach 880980 that was 100,000 feet above the ground in New York City, with skydiving gear, could you survive?
>
> **—Jack Catten**

> **Q** If there was no water on Earth, would we all live?
>
> **—Karen**

These two scenarios are equally unsurvivable.

SCENARIO	SURVIVAL ODDS
RELATIVISTIC SKYDIVING	0.0%
WATER ALL GONE	0.0%

> **Q** Is it possible to make a homemade jet pack?
>
> **—Azhari Zadil**

It's pretty easy to make a jet pack that works once. Twice or more is much harder.

Q I was wondering whether there's a way to use my welder as a defibrillator? (The specific model I own is an Impax IM-ARC140 arc welder.)

—Łukasz Grabowski, Lancaster, UK

You should definitely not use your arc welder as a defibrillator, and after reading your question, I honestly don't think you should be allowed to use it as an arc welder, either.

What if all atoms on Earth were expanded to the size of a grape? Would we survive?

—Jasper

I'm not really sure how to answer this question using science, but now I really want some grapes.

7. T. REX CALORIES

If a T. rex were released in New York City, how many humans/day would it need to consume to get its needed calorie intake?

—T. Schmitz

About half of an adult, or one ten-year-old child.

Shoot, I forgot to eat one yesterday. Am I allowed to double up?

Tyrannosaurus rex weighed about as much as an elephant.*

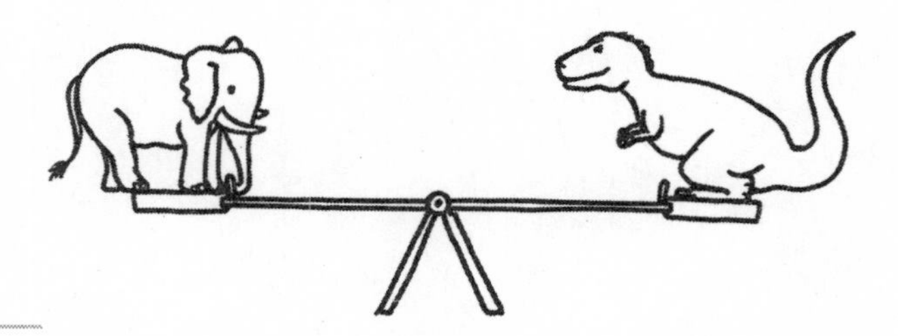

* This always seemed a little off to me; my mental image of elephants is that they're in the same size range as cars or trucks, whereas T. rex, as *Jurassic Park* showed, is big enough to stomp on cars. But a Google image search for car+elephant shows elephants looming over cars just like the T. rex in *Jurassic Park*. So, great, now I'm also afraid of elephants.

No one is totally sure what dinosaur metabolism looked like, but the best guesses for how much food a T. rex ate seem to cluster around 40,000 calories per day.

If we assume dinosaurs had metabolisms similar to today's mammals, they'd eat a lot more than 40,000 calories each day. But the current thinking is that while dinosaurs were more active (loosely speaking, "warm-blooded") than modern snakes and lizards, very large dinosaurs probably had metabolisms that more closely resembled Komodo dragons than elephants and tigers.*

Next, we need to know how many calories are in a human. This number is helpfully provided by Dinosaur Comics author Ryan North, who produced a T-shirt with a human body nutrition label. According to Ryan's shirt, an 80-kg human contains about 110,000 calories of energy, so a T. rex would need to consume a human every two days or so.†

The city of New York had 115,000 births in 2018, which could support a population of about 350 tyrannosaurs. However, this ignores immigration—and, more important, *em*igration, which would probably increase substantially in this scenario.

I'M THINKING OF MOVING OUT OF BROOKLYN.
THE RENT IS SO HIGH, AND EVERYONE
IS GETTING EATEN BY TYRANNOSAURS.

* For big sauropods, we know this must be the case, because if they had metabolisms like mammals they would overheat. However, there's a lot of uncertainty surrounding T.-rex-size dinosaurs.

† A T. rex would likely be willing to eat several days' to weeks' worth of food in one meal, so if it has the option, it might eat a bunch of people at a time, then go for a while without eating.

The 39,000 McDonald's restaurants worldwide sell something like 18 billion hamburger patties per year,* for an average of 1,250 burgers per restaurant per day. Those 1,250 burgers contain about 600,000 calories, which means that each T. rex only needs about 80 hamburgers per day to survive, and one McDonald's could support more than a dozen tyrannosaurs on hamburgers alone.

If you live in New York and you see a T. rex, don't worry. You don't have to choose a friend to sacrifice; just order 80 burgers instead.

And then if the T. rex goes for your friend, anyway, hey, you have 80 burgers.

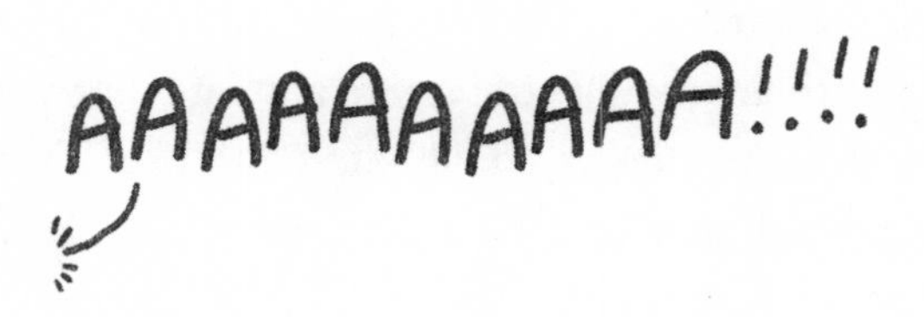

Maybe the friend was more of an acquaintance, anyway.

* They stopped updating the "x billions served" number on their signs in the mid-1990s, so this is just a rough estimate.

8. GEYSER

If one were to stand on top of the Old Faithful geyser in Yellowstone National Park, at what speed would they be launched upward by the water, and what injuries would they likely sustain?

—Catherine McGrath

You would not be the first person to be severely burned by Old Faithful, although you might be the first person to die that way.

The book *Death in Yellowstone*, a chronicle of fatal incidents and accidents in Yellowstone National Park compiled by park historian Lee H. Whittlesey, doesn't mention any deaths from the geyser jets themselves. People were regularly burned by eruptions—including a German surgeon who survived a fall into Old Faithful's vent in 1901—but there are no well-documented cases of deaths from the geyser blast.

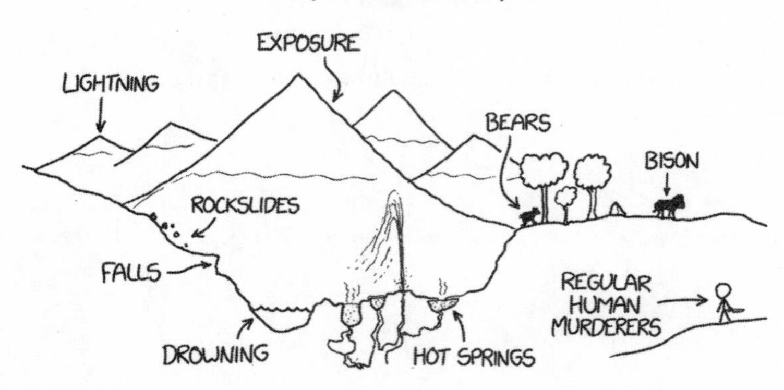

But while *Death in Yellowstone* doesn't mention any deaths from the geyser jets themselves, it recounts an alarming number of incidents *near* them. Often, the boiling pools in geothermally active areas will be covered with a thin, fragile mineral crust. People who walk around the geysers regularly step through the crust and plunge to their deaths.*

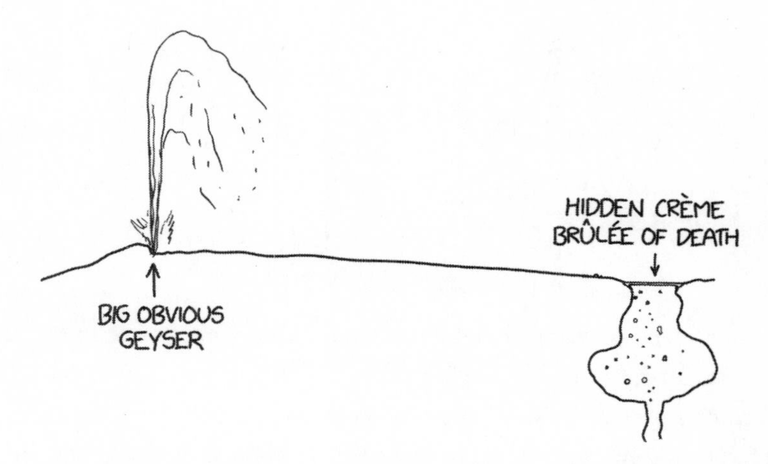

If you did make it to the geyser and stood over it when it erupted, the experience would not be fun.When Old Faithful erupts, it ejects about half a ton of water per second. The jet that emerges is a mix of water droplets, air, and steam with the density of a bag of cotton balls. This jet moves fast—about 70 m/s right before it emerges from the ground—so it carries the momentum of a stream of cars on the highway.

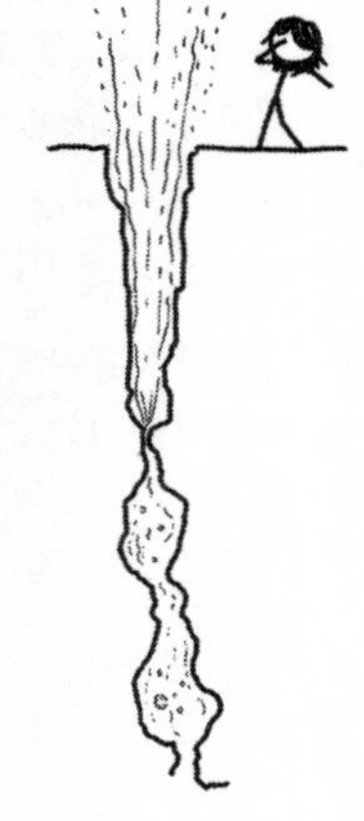

A geyser is sort of like an upside-down rocket. If you calculate Old Faithful's thrust the same way you would a rocket engine, by multiplying the rate of mass flow by its speed, you come up with a few thousand pounds of force. This is similar to the thrust of a fighter jet's ejection seat, which tells us that it's clearly powerful enough to launch a person high into the air.

* In one incident in 1905, the unlucky person was actually taking notes on the geyser in a notebook when she fell in, which I find uncomfortably relatable. I'm pretty sure that's how I'll go.

SIMPLIFIED MODELS

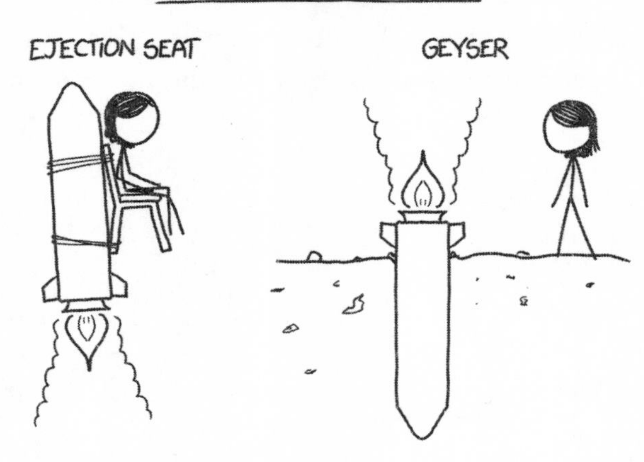

Your launch speed—and the height you would fly—would depend a lot on exactly how the geyser jet hit you. A glancing blow might just knock you off to one side. You could catch more of a lift if you were centered directly over the vent, blocking as much of the stream as possible. If you were holding a very sturdy umbrella, you could in principle launch yourself hundreds of feet into the air, even higher than the plume itself. Even if you survived the severe burns, the landing would almost certainly be fatal.

A surprising number of people have been scalded by Yellowstone's geysers. In the 1920s, about one person a year was burned by Old Faithful. Unlike those who fell into the boiling pools, the people scalded by the geyser generally weren't just people who accidentally wandered onto an unsafe spot without realizing it. Most of them were leaning over and trying to peer down into the steam vent.

I guess we need to add another item to the list.

THINGS YOU SHOULD NOT DO
(PART 3,647 OF ????)

#156,812 EAT TIDE PODS
#156,813 WALK ON STILTS IN A THUNDERSTORM
#156,814 SET OFF FIREWORKS AT A GAS STATION
#156,815 FEED YOUR CAT TREATS THAT ARE THE EXACT SHAPE AND TEXTURE OF A HUMAN HAND
#156,816 (NEW!) LEAN OVER A GEYSER VENT AND TRY TO LOOK DOWN INTO IT

9. PEW, PEW, PEW

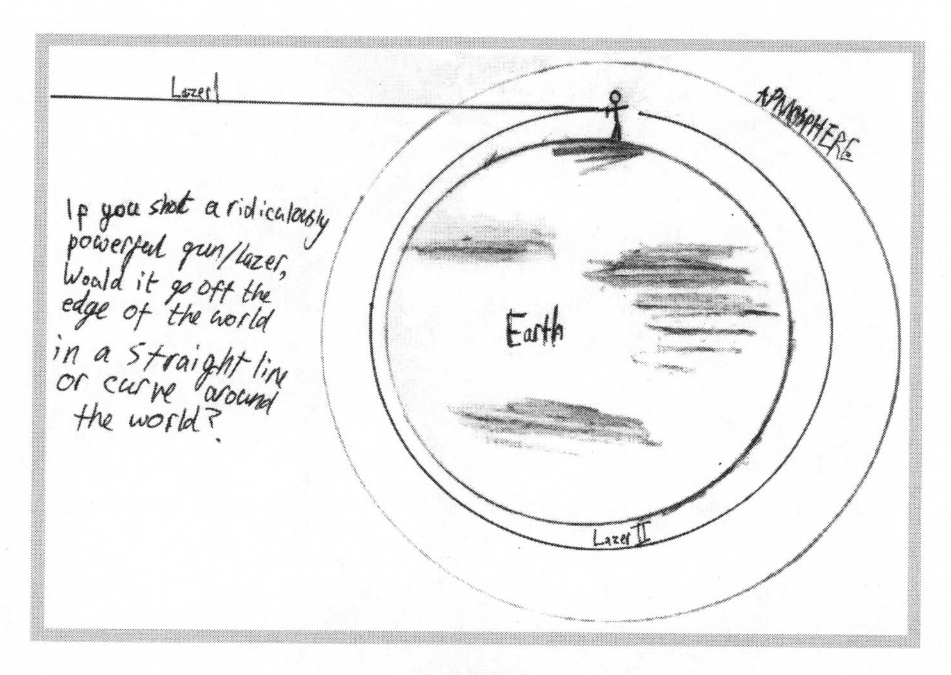

—Maelor, age 11

Path #1 is right. The beam would go off the edge of the Earth into space! Probably.

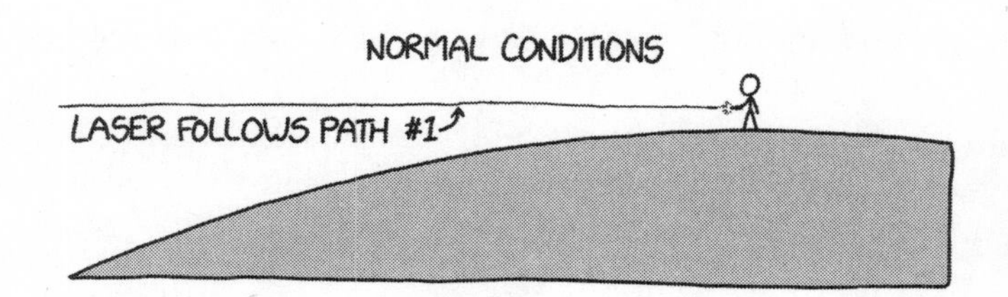

There are a few rare cases where the beam *won't* go off the edge of the Earth. If you stand near an ocean on a hot day, at just the right time and place, you can get a laser to start to follow path #2 instead.

Light in the atmosphere doesn't go in a perfectly straight line. Air slows light down, and denser air slows the light down more. When the air on one side of the beam is being slowed down more than the other side, it makes the light bend in that direction.

In most of the atmosphere, light bends slightly downward, because the air below it is much denser than the air above it.*

Near the ground, you often find layers of air with very different temperatures close together. On hot, sunny days, the ground can get hot, which makes the air right near the ground get hot, too. This is why when you look at a parking lot, you

* The atmosphere bends the sun's light, too. At sunrise, when you see the sun appear, it's actually still a little below the horizon. If we didn't have an atmosphere, you wouldn't see it. But the atmosphere bends the light, letting you see it a little bit early.

sometimes see what looks like shimmering water—a mirage. A mirage is a reflection of the sky; light from the sky comes down near the surface and then bends up toward your eye, so it looks like it's coming from the ground.

If you aimed a laser gun at that patch of "water," it would bend up and sail off into the sky.

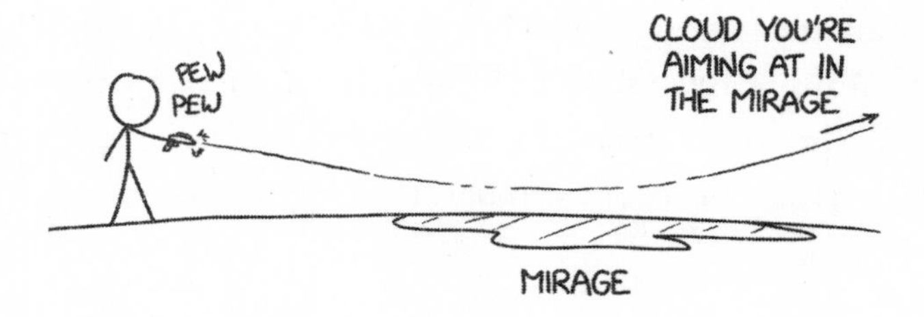

If you want to bend the laser enough to keep it from going off to space, you need to find a place where the air near the ground is *colder* than the air right above it. One place where this happens is over the ocean: When hot air goes over cold ocean water, it cools down the air at the surface, like a parking lot in reverse. Light passing over the cold air bends down, sometimes by a lot.

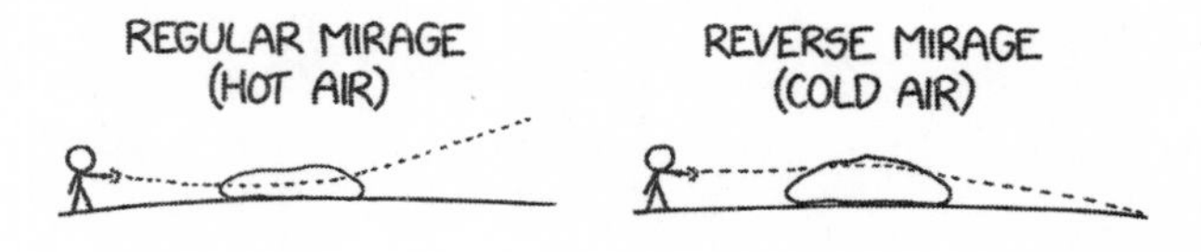

When you look over the water, you sometimes see land and water floating above the surface, because of the funny paths the light takes. These shimmering clumps of land and buildings floating above the horizon are called a Fata Morgana, so named because people thought they looked like the floating castles of the sorceress Morgan le Fay.

I SEE A FLOATING ISLAND! IT MUST BE THE CASTLE OF THE EVIL SORCERESS MORGAN LE FAY!
I THINK THAT'S JUST THE OTHER SIDE OF THE BAY. SEE, THAT'S MEGAN'S HOUSE.
MEGAN LE FAY!!

If you want to shoot a laser at a Fata Morgana, just aim straight at it. It's not really there, but the path the laser takes will be the same one the light reaching your eyes takes. The thing floating in the sky is an illusion, but illusions are made of light. So, if you're ever faced with some kind of frightening illusory phantom, just remember this handy optics rule: If you can see it, you can shoot it with a laser.

PEW
PEW
PEW
WILL YOU STOP THAT!?

10. READING EVERY BOOK

At what point in human history were there too many (English) books to be able to read them all in one lifetime?

—Gregory Willmot

This is a complicated question. Getting accurate counts of the number of extant books at different times in history is very hard, bordering on impossible. For example, when the Library of Alexandria burned, a lot of writing was lost,* but *how much* writing was lost is hard to pin down. Some estimates range from 40,000 books to 532,800 scrolls. Other writers call those numbers implausible in one direction or the other.

Researchers Eltjo Buringh and Jan Luiten van Zanden used historical book catalogs to put together statistics on the number of books (or manuscripts) published annually per region. By their figures, the rate of publication in the British Isles probably passed one manuscript per day in around the year 1075 CE.

Most of the manuscripts published in 1075 weren't in English, or even in the variants of English common at the time. In 1075, literature in Great Britain was typically written in some form of Latin or French, even in areas where Old English was commonly spoken on the street.

The stories comprising *The Canterbury Tales* (written in the late 1300s) were part of a move toward vernacular English as a literary language. While they're technically written in English, they're not exactly readable to a modern eye:

* On the other hand, a lot of Egyptian readers were probably excited to get out of overdue book fines.

> *"Wepyng and waylyng, care and oother sorwe*
> *I knowe ynogh, on even and a-morwe,"*
> *Quod the Marchant, "and so doon other mo*
> *That wedded been."*

(If my ninth-grade English teacher is reading this, don't worry, I'm just making a joke. I totally understand that passage.)

Even if we know how many manuscripts were published per year, in order to answer Gregory's question, we need to know how long it takes to *read* a manuscript.

Rather than trying to figure out how long all the lost books and codices are, we can step back and take a longer view of things.

WRITING SPEED

Tolkien wrote *The Lord of* *the Rings* in 11 years, which means that he wrote at an average pace of 125 words per day, or less than 0.085 word per minute. Harper Lee wrote the 100,000-word *To Kill a Mockingbird* in two and a half years, for an average of 100 words per day, or 0.075 words per minute. Since *To Kill a Mockingbird* is her only published book, her lifetime average is 0.002 words per minute, or about 3 words per day.

Some writers are substantially faster. Author Corín Tellado published thousands of romance novels in the mid to late twentieth century, turning in a book a week to her publisher. For much of her career, she published well over a million words per year, giving her an average of 2 words per minute over most of her lifetime.

It's reasonable to assume historical writers had a similar range of speeds. You might point out that typing on a keyboard is more than twice as fast as writing a manuscript in longhand. But typing speed isn't a writer's bottleneck. After all, at a typing speed of 70 words per minute, it should only take 24 hours to type out *To Kill a Mockingbird*.

Typing and writing speeds are so different because the bottleneck in writing books is how quickly our brains can organize, produce, and edit stories. This "storytelling speed" has probably changed much less over time than our physical writing speed has.

This gives us a much better way to estimate when the number of books became too large to read. If the average living writer, over their entire lifetime, falls somewhere between Harper Lee and Corín Tellado, they might produce 0.05 words per minute during their lifetime.

The average person can read at 200 to 300 words per minute. If you were to read for 16 hours a day at 300 words per minute, you could keep up with a world containing an average population of 100,000 living Harper Lees or 200 living Corín Tellados.

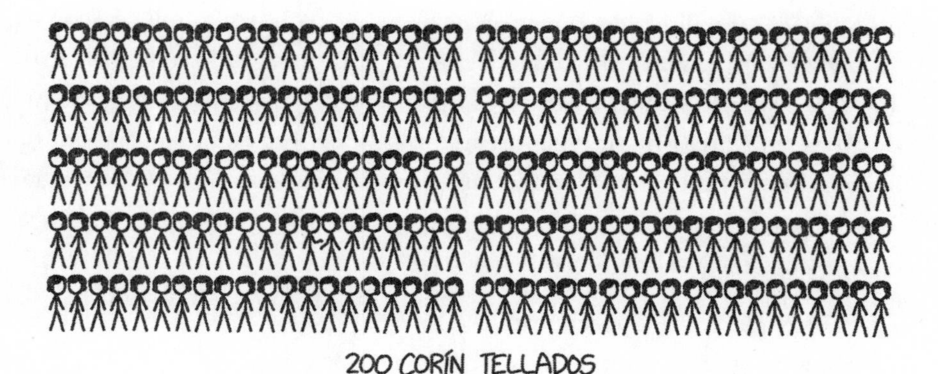

If we estimate that during their *active* periods, writers are producing somewhere between 0.1 and 1 word per minute, then one dedicated reader might be able to keep up with a population of about 500 or 1,000 active writers. The answer to Gregory's question—the date at which there were too many English books to read in a lifetime—happened sometime before the population of active English writers reached a few hundred. At that point, catching up became impossible.

The magazine *Seed* estimates that the total number of authors reached this point around the year 1500 and has continued rising rapidly ever since. The number of active English writers crossed this threshold shortly thereafter, around the time of Shakespeare, and the total number of books in English probably passed the lifetime reading limit sometime in the late 1500s.

On the other hand, how many of them would you *want* to read? If you go to goodreads.com/book/random, you can see a semirandom sample of what you'd be reading. Here's what came up for me:

- *School Decentralization in the Context of Globalizing Governance: International Comparison of Grassroots Responses*, by Holger Daun
- *Powołanie* (Dragon Age #2), by David Gaider
- *An Introduction to Vegetation Analysis: Principles, Practice and Interpretation*, by David R. Causton
- *AACN Essentials of Critical-Care Nursing Pocket Handbook*, by Marianne Chulay
- *National righteousness and national sin: the substance of a discourse delivered in the Presbyterian church of South Salem, Westchester co., N.Y., November 20, 1856*, by Aaron Ladner Lindsley
- *Phantom of the Auditorium* (Goosebumps #24), by R. L. Stine
- *High Court #153; Case Summaries on Debtors and Creditors-Keyed to Warren*, by Dana L. Blatt
- *Suddenly No More Time*, by Emil Gaverluk

So far, I've read . . . the Goosebumps book.

To make it through the rest, I might need to recruit some help.

weird & worrying #1

Q Can bees or other animals go to hell? Or can they murder other bees without consequences?

—Sadie Kim

BEE-LZEBUB

Q How many mirrors reflecting (sun)light would it take to kill or, at least, injure somebody?

—Eli Collinge

MIRROR, MIRROR, ON THE WALL,
I HAVE A FAVOR TO ASK.

Q If you had to remove the tonsils of a giant, what would be the safest way for you to do it? The surgeon is a normal human.

—Tirzah, age 10

HI, I'M A NORMAL HUMAN!
I...DON'T THINK YOU ARE.

Q What would it take to defeat *Air Force One* with a drone???

—Anonymous

HELLO, SECRET SERVICE? YES, IT'S RANDALL AGAIN...

11. BANANA CHURCH

Can all the world's bananas fit inside all of the world's churches? My friends have had this argument for a little over 10 years now.

—Jonas

Yes.

We know the bananas can fit for the simple reason that the world's *people* can probably fit into the world's houses of worship, and people don't go through their own weight in bananas each year.

According to a 2017 Pew Research survey of religious observance,* slightly less than 30 percent of the world's population attends weekly services as part of their religious tradition. If we count all the spaces where those services happen as "churches," it suggests that there are enough of these spaces to accommodate at least 2 billion people.

* With some guesswork to fill in the gaps for countries not covered by their surveys.

Buildings like churches and classrooms generally have between 5 and 25 square feet of floor space per person. If we suppose that they average 15 square feet per person, and that most of them serve only one congregation, then it means that places of worship occupy something like 1,000 square miles of the Earth's surface.

Let's say we could gather up an entire year's supply of bananas, which is probably about 120 million tons. When packed into boxes, bananas have a density of about 300 kilograms per cubic meter. To see how deep they would fill the world's houses of worship, we can divide their total volume by our 1,000-square-mile estimate:

$$\frac{120 \text{ million tons}}{16 \text{ kg}/(10 \text{ in} \times 16 \text{ in} \times 20 \text{ in})} / 1000 \text{ miles}^2 = 6 \text{ inches}$$

The result tells us that a year's supply of bananas would only come up to a person's ankles.

The layer of bananas would be even shallower than that, because a year's supply of bananas doesn't actually exist at any one time. Banana flowers take a few months to ripen from small fingerlike fruits to full-size ripe bananas ready to be eaten . . .

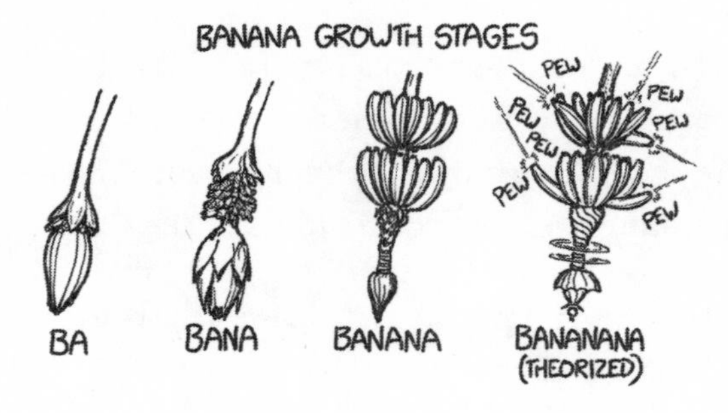

. . . so the number of bananas in existence at any given time is a fraction of a year's production, making the banana layer even shallower.

Even if our banana data is wrong, the answer is still probably right. Flipping the calculation around, we can calculate how many bananas would be needed to fill all of the world's churches, and see if that number seems plausible.

If about 1 in 4 people attend regular indoor worship services, and the buildings in which they do so have about 15 square feet of space per attendee, then that's about 4 square feet worth of floor space for each person on Earth (including nonattendees). If there *were* enough bananas in the world to fill all of the houses of worship to their ceilings, that would mean that each person's share of global banana production would fill a volume 2 feet by 2 feet times the average ceiling height.

Many religious buildings are famous for their high ceilings. But even if we assume the average ceiling height is a relatively low eight feet, filling up each person's space would require about 2,000 bananas. I'm pretty sure the world doesn't produce 2,000 bananas per person in a year, for the simple reason that I don't eat six bananas a day and I don't know anyone else who does, either.

Unless there's one person out there who eats so many bananas that they throw off the global average.

BANANAS GEORG, WHO LIVES
ON A MOUNTAIN AND EATS 17
TRILLION BANANAS PER YEAR,
IS AN OUTLIER AND SHOULD
NOT HAVE BEEN COUNTED

12. CATCH!

Is there any way to fire a gun so that the bullet flies through the air and can then be safely caught by hand? E.g., the shooter is at sea level and the catcher is up on a mountain at the extreme range of the gun.

—Edmond Hui, London

The "bullet catch" is a stage trick in which a performer appears to catch a fired bullet in midflight—often between their teeth. This an illusion, of course; it's not possible to catch a bullet like that.

But under the right conditions, you *could* catch a bullet. It would just take a lot of patience and luck.

A bullet fired straight up would eventually reach a maximum height.* It probably wouldn't stop completely; more likely, it would be drifting sideways at a couple of meters per second. If someone fired a bullet upward . . .

---

* Don't do this. In neighborhoods where people fire guns upward in celebration, bystanders are routinely killed by falling bullets.

. . . and you were hanging out in a hot-air balloon directly above the firing range . . .

. . . it's *possible* that you could reach out and grab the bullet at the apex of its flight.

THINGS YOU SHOULD NOT DO
(UPDATED LIST)

#156,812 EAT TIDE PODS
#156,813 WALK ON STILTS IN A THUNDERSTORM
#156,814 SET OFF FIREWORKS AT A GAS STATION
#156,815 FEED YOUR CAT TREATS THAT ARE THE EXACT SHAPE AND TEXTURE OF A HUMAN HAND
#156,816 LEAN OVER A GEYSER VENT AND TRY TO LOOK DOWN INTO IT
#156,817 (NEW!) FLY A HOT-AIR BALLOON THROUGH A FIRING RANGE

If you succeeded at grabbing a bullet at the peak of its arc, you might notice something odd: In addition to being hot, the bullet would be spinning. It would have lost its upward momentum but not its rotational momentum; it would still have the spin given to it by the barrel of the gun.

This effect can be seen, dramatically, when a bullet is fired at ice. As confirmed by dozens of YouTube videos, bullets fired into ice are often found still spinning rapidly. You'd have to grab the bullet firmly; otherwise, it might jump out of your hand.

If you don't have a hot-air balloon, you could potentially make this work from a mountain peak. Mount Thor* in Canada features a vertical drop of 1,250 meters. According to ballistics lab Close Focus Research, this is almost exactly how high a .22 long rifle bullet will fly if fired directly upward.

* Which we previously encountered in the Free Fall chapter of the first *What If?* book.

If you want to use larger bullets, you'll need a much larger drop; an AK-47's bullet can go more than 2 kilometers upward. Earth doesn't have any purely vertical cliffs that tall, so you'd need to fire the bullet at an angle, and it would have significant sideways speed at the top of its arc. However, a suitably tough baseball glove might be able to snag it.*

In any of these scenarios, you'd have to get extraordinarily lucky. Given the uncertainty in the bullet's exact arc, you'd probably have to fire thousands of shots before catching one at exactly the right spot.

And by that point, you may find you've attracted some attention.

* In fact, according to *Rifle* magazine, a gun writer once claimed that at a thousand yards, he could catch ordinary rifle bullets with a baseball glove. Of course, he was being figurative—you wouldn't see the bullet coming, so you'd be just as likely to catch it with your face as with your glove.

13. LOSE WEIGHT THE SLOW AND INCREDIBLY DIFFICULT WAY

I want to lose 20 pounds. How much of the Earth's mass would I have to "relocate" to space in order to achieve my goal?

—Ryan Murphy, New Jersey

This seems simple enough. Your weight comes from the Earth's gravity pulling you down. The Earth's gravity comes from its mass. Less mass should mean less gravity. Remove mass from the Earth, and you'll lose weight.

You decide to give it a try.

Removing lots of mass from the Earth will take a lot of energy, so you start by seizing the entire planet's oil reserves.

You process the oil into fuel and use it to launch several hundred billion tons of rocks into orbit. This shaves off an average of 0.2mm of rock from the Earth's surface. You hop on the scale.

Okay, it didn't work. But that makes sense; a few hundred billion tons is a tiny fraction of the Earth's mass.

Burning the Earth's other fossil fuels helps a little—especially coal, which there's quite a lot of—and lets you remove almost a millimeter of the Earth's surface.* You step back on the scale.

Darn.

You need more energy.

You cover the entire planet with highly efficient solar panels and spend a year soaking up all the sunlight that strikes the Earth and using it to power your rock launchers. Humanity lives in the shade under your panels. People are probably pretty mad at you at this point.

* People might complain, but on the plus side, that millimeter probably includes all the grime and dirt on the floor. Maybe you can spin it as a free cleaning.

A year's worth of sunlight would give you enough energy to remove nearly 100 trillion tons of rock—several inches of the planet's surface. Sadly, that's not enough.

Clearly, this incremental approach isn't working.

You need more power. Rather than capturing only the small portion of the Sun's energy that hits Earth, you decide to capture *all* of its energy by building an energy-collecting enclosure around it—a Dyson sphere. Once you've harnessed the Sun's entire output, you have enough energy to start stripping away the Earth's surface much more quickly.

The Earth's rocks get hotter the deeper you go. After you strip away a few hundred meters of the crust, people start to notice that the ground is warming up. By the time you remove a kilometer of rock, the surface is up to 40°C. That might feel nice on your feet when you get out of bed on a cold morning, but it will make life pretty uncomfortable. Also, since you've removed the tops of all the various hot spots, all the world's volcanoes would erupt.

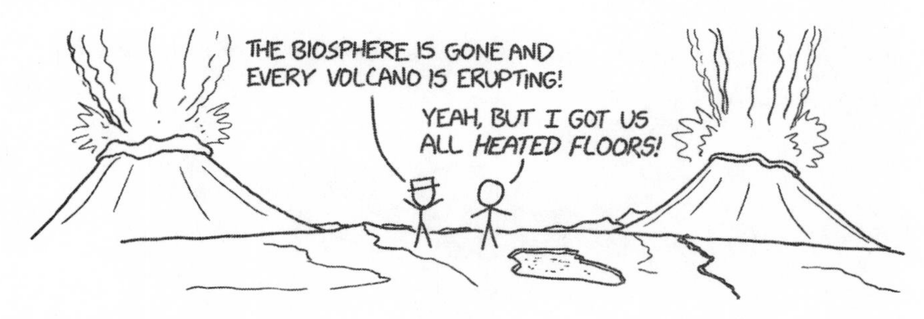

You check the scale.

Darn.

You use your Dyson sphere to remove more rock. You've now stripped away a 5-kilometer layer, which takes about 20 minutes. (For good measure, you spend another few minutes removing the oceans.) The Earth is no longer remotely habitable. Thanks to the exposed magma under the Yellowstone supervolcano, northwestern Wyoming is a lake of lava. The ground in most places is hot enough to boil water and start fires.

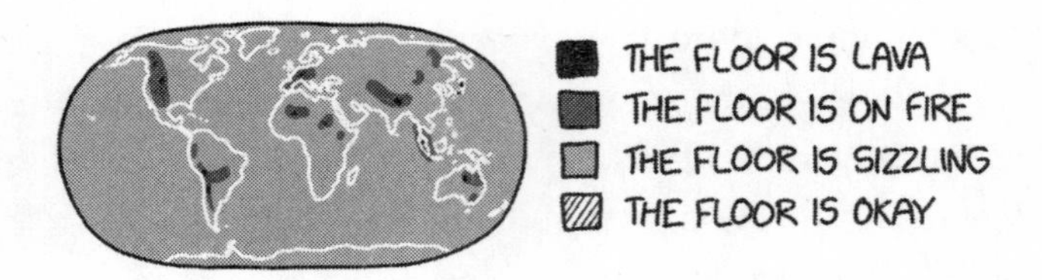

You try the scale again.

That's fine, you just need to remove more rock, perhaps with some kind of Sun-powered vegetable peeler.

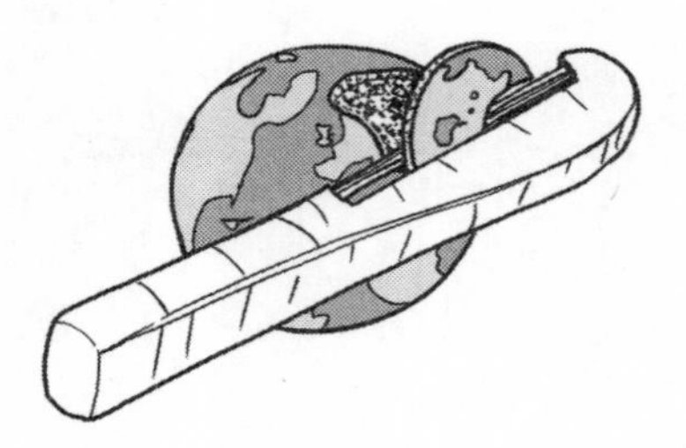

You slice away 20 kilometers of crust, which exposes the Earth's mantle over much of the former sea floor.

Well, no one ever said losing weight was easy. You take off another 20 kilometers, removing layers of molten mantle and pockets of deep crust.

You keep going. After four hours of work with your planet peeler, you've removed 60 kilometers of mostly molten rock. When you step on the scale, you finally see a change.

You're one pound *heavier.*

How could this be?

If the Earth were of uniform density, removing layers would make you lighter. But our planet gets denser the deeper you go, and the density cancels out the mass loss. The planet is getting a little lighter as you remove the surface, but you're also getting closer to that dense core. The net effect is that removing the Earth's outer layer makes its surface gravity *stronger.*

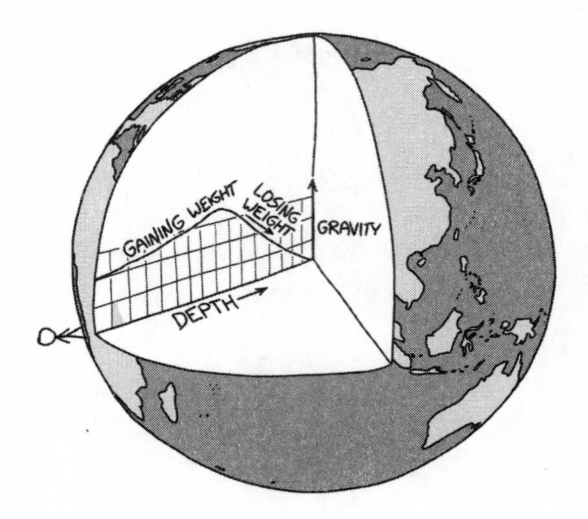

Gravity keeps increasing as you go deeper. It only levels off after you've shaved off about 3,000 kilometers, reducing the Earth's diameter by half and ejecting two-thirds of its mass. (This takes your sun-powered planet peeler about a week.) Your weight peaks at about 207 pounds, after which it starts falling as you start removing the denser outer core.

Once you've removed 3,450 kilometers of rock, your weight gets back down to what it was when you started. After 3,750 kilometers of rock, you finally achieve your goal of losing 20 pounds. At this point, you've removed 85 percent of the Earth's mass. But you've lost weight!

This plan has some flaws. It destroys the Earth, yes, but it's also unnecessarily inefficient. There's a much easier way to reduce the Earth's gravitational pull on you without changing your mass or leaving the surface.

A spherical shell of matter doesn't exert any gravitational force on objects inside it, which means that if you go underground, the layers of rock above you stop contributing to your weight. From a gravitational point of view, it's as if they vanish. You didn't actually need to *remove* mass from the Earth, you just needed to go under it. You could've avoided all that work with a comparatively simple tunnel.

Did you at least avoid exercise? Well, sort of. Your project ended up requiring you to do an awful lot of work. Removing the Earth's surface required 5×10^{28} calories of energy, which is more calories than would be burned if the entire human population started doing intense workouts 24 hours a day from now until when the sun burned out and its remnant cooled to room temperature.

	WORK REQUIRED (CALORIES BURNED)
YOUR PLAN	50,000,000,000,000,000,000,000,000,000
EVERY OTHER PLAN ANYONE HAS EVER COME UP WITH	LESS THAN THAT

If your goal was to avoid work, you could not have failed more badly.

14. PAINT THE EARTH

Has humanity produced enough paint to cover the entire land area of the Earth?

—Josh, Woonsocket, RI

This answer is pretty straightforward to calculate. We can look up the size of the world's paint industry, extrapolate backward to figure out the total amount of paint produced, and make some assumptions about how we're painting the ground.*

But first, let's think about different ways we might come up with a guess about what the answer will be. In this kind of thinking—often called Fermi estimation—all that matters is getting in the right ballpark; that is, the answer should have about the right number of digits. In Fermi estimation, you can round† all your answers to the nearest order of magnitude:

* When you get to the Sahara Desert, I recommend not using a brush.

† Using the formula $Fermi(x) = 10^{round(log_{10}x)}$, meaning that 3 rounds to 1 and 4 rounds to 10.

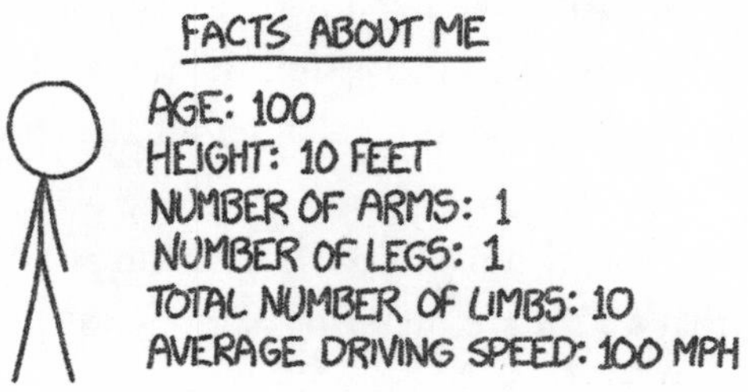

FACTS ABOUT ME

AGE: 100
HEIGHT: 10 FEET
NUMBER OF ARMS: 1
NUMBER OF LEGS: 1
TOTAL NUMBER OF LIMBS: 10
AVERAGE DRIVING SPEED: 100 MPH

Let's suppose that, on average, everyone in the world is responsible for the existence of two rooms, and they're both painted. My living room has about 50 square meters of paintable area, and two of those would be 100 square meters. Eight billion people times 100 square meters per person is a little under a trillion square meters—an area smaller than Egypt.

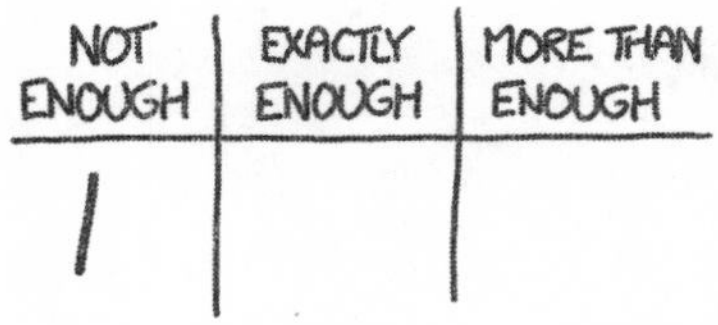

NOT ENOUGH	EXACTLY ENOUGH	MORE THAN ENOUGH
/		

Let's make a wild guess that, on average, one person out of every thousand spends their working life painting things. If I assume it would take me three hours to paint the room I'm in,* and 100 billion people have ever lived, and each of them spent 30 years painting things for 8 hours a day, we come up with 150 trillion square meters . . . just about exactly the land area of the Earth.

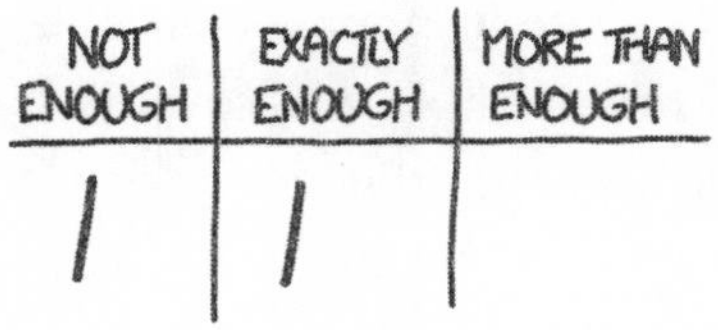

NOT ENOUGH	EXACTLY ENOUGH	MORE THAN ENOUGH
/	/	

How much paint does it take to paint a house? I'm not enough of an adult to have any idea, so let's take another Fermi guess.

* This is probably optimistic, especially if there's an internet connection in the room.

Based on my impressions from walking down the aisles, home improvement stores stock about as many lightbulbs as cans of paint. A normal house might have about 20 lightbulbs, so let's assume that a house needs about 20 gallons of paint.[*] Sure, that sounds about right.

The average US home costs about $400,000. Assuming each gallon of paint covers about 300 square feet, that's a square meter of paint per $70 of real estate. I vaguely remember that the world's real estate has a combined value of something like $400 trillion,[†] which suggests that there's about 6 trillion square meters of paint on the world's real estate. That's a little less than the area of Australia.

NOT ENOUGH	EXACTLY ENOUGH	MORE THAN ENOUGH
II	I	

Of course, both of the building-related guesses could be overestimates (lots of buildings are not painted) or underestimates (lots of things that are not buildings[‡] are painted). But from these wild Fermi estimates, my guess would be that there probably isn't enough paint to cover all the land.

So, how did Fermi do?

According to the *Polymers Paint Color Journal*, the world produced 41.5 billion liters of paints and coatings in 2020.

There's a neat trick that can help us here. If some quantity—say, the world economy—has been growing for a while at an annual rate of n—say, 3 percent (0.03)—then the most recent year's share of the whole total so far is $1 - \frac{1}{1+n}$ and the whole total so far is the most recent year's amount times $1 + \frac{1}{n}$.

---

* These are very rough estimates.

† Citation: this really boring dream I had once.

‡ EXAMPLES OF THINGS THAT ARE NOT BUILDINGS: ducks, leaves, M&Ms, cars, the Sun, sand grains, cuttlefish, microchips, nail polish remover, the moons of Jupiter, lightning, mouse fur, zeppelins, tapeworms, pickle jars, those sticks you use to toast marshmallows, alligators, tuning forks, minotaurs, Perseid meteors, ballots, crude oil, social media influencers, and catapults that throw handfuls of engagement rings. Those are all the nonbuildings I can think of; if you can think of anything I missed, you can make a note of it here in the margin.

If we assume paint production has, in recent decades, followed the economy and grown at about 3 percent per year, that means the total amount of paint produced equals the current yearly production times 34.* That comes out to about 1.4 trillion liters of paint. At 30 square meters per gallon,† that's enough to cover 11 trillion square meters—less than the area of Russia.

So the answer is no; there's not enough paint to cover the Earth's land, and—at this rate—there probably won't be enough until the year 2100.

Score one for Fermi estimation.

ENRICO FERMI'S FAVORITE MOVIES:

- 100 DALMATIANS
- OCEAN'S TEN
- T10N
- 1000: A SPACE ODYSSEY
- MIRACLE ON 100TH ST
- THE TENTH SENSE
- 10 MILE
- THE 100-YEAR-OLD VIRGIN

* $1 + \frac{1}{0.03}$

† "Square meters per gallon" is a pretty obnoxious nonmetric unit, but it could be worse. I've encountered *acre-foot*, a unit of volume equal to one foot times one chain times one furlong, in actual technical papers.

15. JUPITER COMES TO TOWN

Dear Randall, what would happen if you shrunk Jupiter down to the size of a house and placed it in a neighborhood, say, replacing a house?

—Zachary, age 9

This is one of those questions that sounds like it's going to create a disaster, but when you think about it for a moment, it actually doesn't seem like it would be so bad. Then, if you think about it a little more, you realize it would be *extremely* bad.

A house-size Jupiter wouldn't have much gravity, so it wouldn't create a black hole or anything.* Jupiter is only a little denser than water, so a 50-foot-wide Jupiter would only weigh about 2,500 tons. That's heavy, but it's not *that* heavy; it's as much as a small office building or a few dozen whales. If you put a 50-foot sphere of water in the middle of your neighborhood, it would create a huge mess and might destroy nearby houses before forming a small pond, but it wouldn't do any weird gravity stuff.

Since Zachary's Jupiter is only about the size and weight of a 50-foot sphere of water, it seems like it might not be so bad.

Here's the problem: Jupiter is *hot*.

Like Earth, Jupiter consists of a thin, cool outer layer wrapped around a blisteringly hot interior. Jupiter's interior is mostly hydrogen, compressed and heated to tens of thousands of degrees. And hot, dense things want to expand.

A ball of hydrogen at 20,000°C would push outward with incredible pressure. The reason the actual Jupiter doesn't explode is because its massive gravity counteracts that pressure, holding it together. If you shrink Jupiter and plop it down in the middle of your neighborhood, that hot, high-pressure hydrogen, with no gravity holding it together, would expand.

* We're assuming the smaller Jupiter's density stays the same—it's made of the same stuff, there's just less of it. These are *Honey, I Shrunk the Kids* rules.

Jupiter would expand so violently that it would flatten all the houses on your block almost instantly, and probably take the whole neighborhood with it. As the fireball grew, it would cool, and rise into the atmosphere. After five or ten seconds the rising gas would form a mushroom cloud.*

If you recorded these events—hopefully from a safe distance—and played the video in reverse, it would, in a way, resemble Jupiter's formation.

The *reason* Jupiter is so hot is that 4.6 billion years ago gravity caused a cloud of gas to collapse together. When you compress gas, it heats up, because the molecules are getting smashed together and bouncing around faster. Since a lot of gas fell together to form Jupiter, its gravity was very strong, so it pulled itself together hard and got extremely hot.

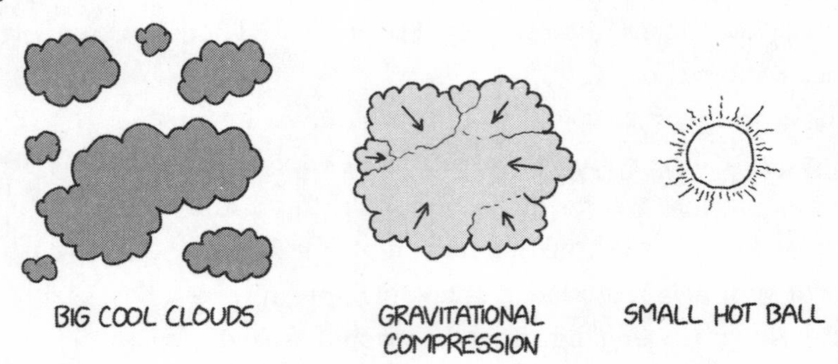

* We associate mushroom clouds with nuclear weapons, but really, they're just what happens when you dump a lot of heat energy into the air all at once. It doesn't really matter what the source of the heat is—if there's enough of it and it's released fast enough, it will create a mushroom cloud.

Over four billion years later, a lot of that heat—about half of it—is still there, trapped under Jupiter's immense gravity and insulating blanket of clouds. A mini-Jupiter would lack that crushing inward pull. Its hot core would be able to throw off its insulating blanket and expand outward, spreading out and rapidly cooling.

The neighborhood-destroying blast would represent 4 billion years of pent-up heat finally being released. Jupiter, freed from gravity's shackles, would once again become what it was before the Sun formed—a thin, cool cloud of gas, spread out across the sky.

16. STAR SAND

If you made a beach using grains the proportionate size of the stars in the Milky Way, what would that beach look like?

—Jeff Wartes

Sand is interesting.[citation needed]

"Are there more grains of sand than stars in the sky?" is a popular question that has been tackled by many people. The short answer to that question is that there are probably more stars in the visible universe than grains of sand on all of Earth's beaches.

When people try to answer the question of whether there are more stars than sand grains, they often dig up some good data on the number of stars, then do some hand-waving about sand grain size to come up with the equivalent number for sand. Arguably, this is because geology and soil science are more complicated than astrophysics.

We're not going to try to count the sand grains, but to answer Jeff's question, we *do* need to figure out what the deal with sand is. Specifically, we need to have some idea of what grain sizes correspond to clay, silt, fine sand, coarse sand, and gravel, so we can understand how our galaxy would look and feel if it were a beach.*

Fortunately, there's nothing scientists like more than coming up with definitions for categories. A century ago, a geologist named Chester K. Wentworth published a definitive index of grain sizes, which defined size ranges for coarse sand, fine sand,

* Instead of just containing a bunch of them.

and clay. According to surveys of sand, the grains found on beaches tend to run from 0.2mm to 0.5mm (with the finest layers on top). This corresponds to medium-to-coarse sand on Wentworth's scale.

The individual sand grains are about this big:

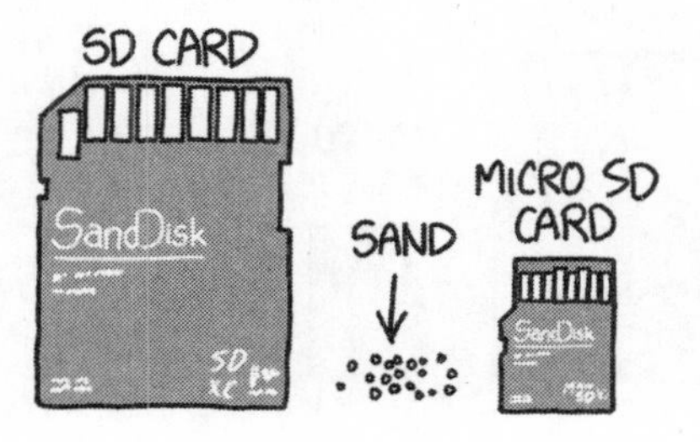

If we assume the Sun corresponds to a typical sand grain, then multiply by the number of stars in the galaxy, we come up with a large sandbox worth of sand.*

If all stars were the same size as the Sun, this estimate would be right, but they're not. Some stars are small and some are huge. The smallest ones are about the size of Jupiter, but some of the big ones are staggeringly large, comparable in size to our whole Solar System. Some of the grains in our sandbox universe would be more like boulders.

Here's how the main-sequence† star-sand grains would look:

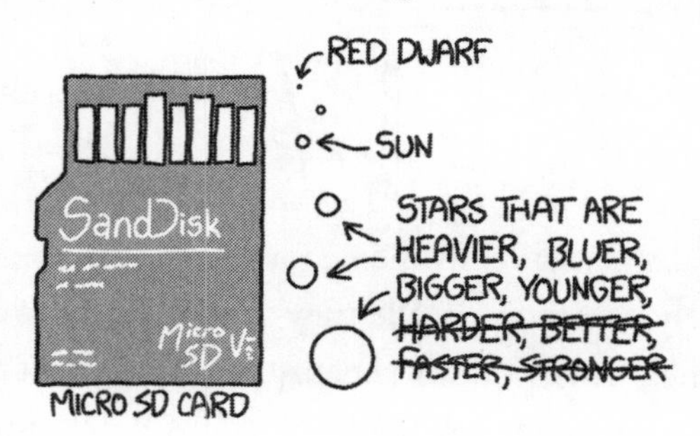

*Astronomy fact: These stars are **all** technically called "dwarf" stars, even the big ones, because astronomers are not as good at naming things as Chester K. Wentworth.*

---

* I mean, we come up with a bunch of numbers, but our imagination turns them into a sandbox.

† The stars in the main part of their fuel-burning life cycle.

The sand versions of these main-sequence stars mostly fall into the "sand" category, though the larger Daft Punk stars cross the line into "granules" or "small pebbles."

However, that's just the main-sequence stars. Dying stars get much, much bigger.

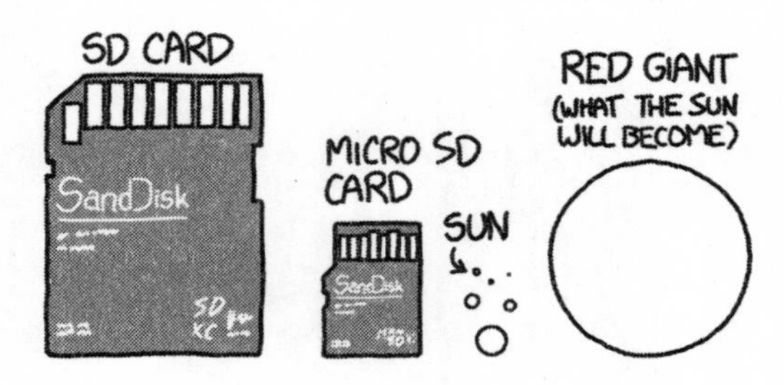

They're almost as big as SD cards!

When a star runs out of fuel, it expands into a red giant. Even ordinary stars can grow to huge sizes, but when a star that's already massive enters this phase, it can become a true monster. These red supergiants are the largest stars in the universe.

These beach-ball-size sand stars would be rare, but the grape-size and baseball-size red giants are relatively common. While they're not nearly as abundant as Sun-like stars or red dwarfs, their huge volume means that they'd constitute the bulk of our sand. We would have a large sandbox worth of grains . . . along with a field of gravel that went on for miles.

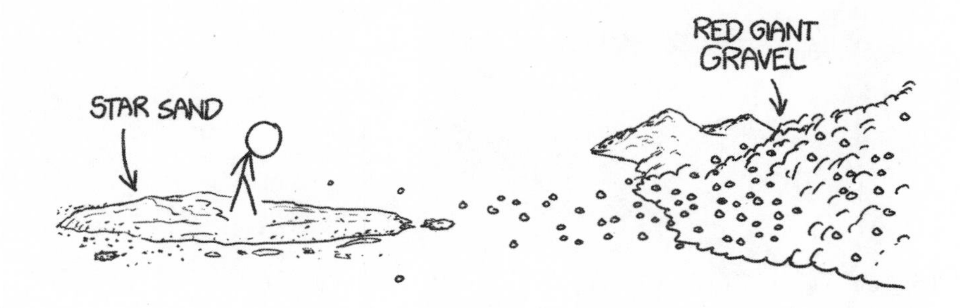

The little sand patch would contain 99 percent of the pile's individual grains, but less than 1 percent of its total volume. Our Sun isn't a grain of sand on a soft galactic beach; instead, the Milky Way is a field of boulders with some sand in between.

But, as with the real Earth seashore, it's the rare little stretches of sand between the rocks where all the fun seems to happen.

17. SWING SET

How tall can a swing set be while still being powered by a human pumping their legs? Is it possible to build a swing set tall enough to launch the rider into space if they jump at the right time? (Assuming the human has enough energy, which my 5-year-old seems to have.)

—Joe Coyle

There's a surprisingly large amount of research into the physics of swing sets, partly because pendulums are really interesting physical systems and probably also because all physicists were once children.

Children who play on swing sets quickly learn that they can get themselves going by pumping—kicking out their feet and leaning back, then tucking in their feet and leaning forward. Physicists call this "driven oscillation," and a series of studies since the 1970s have analyzed exactly how pumping a swing works and what the most efficient way to do it is.

What the physicists found, after half a century of research, was that children know exactly what they're doing. Rhythmically kicking and leaning with their hands on the chains seems to be just about the optimal strategy for powering a swing using the rider's body. For a while, some physicists theorized that a better strategy for pumping a swing might be to stand on the seat and raise and lower your body, by alternately crouching and standing upright, but further calculation showed that the kids have it figured out.

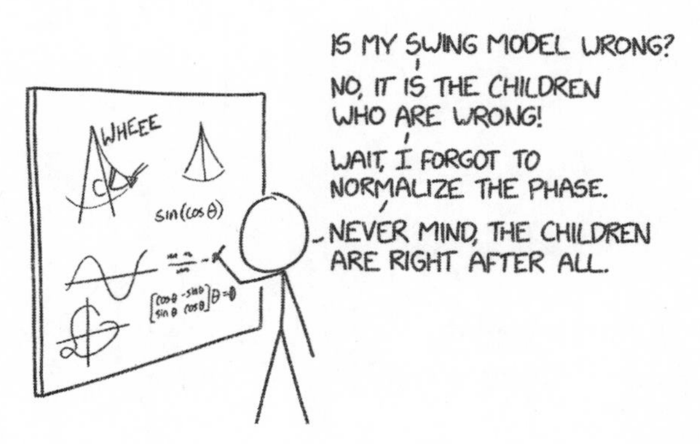

It can seem like pumping your feet to swing higher must violate conservation of energy somehow. How can you push against nothing? But you're not pushing against nothing; you're pushing, indirectly, against the crossbar of the swing set.

If you attach a motorized wheel to the bottom of a pendulum, when you turn on the motor to spin the wheel, the pendulum twists a little in the opposite direction, keeping the angular momentum of the whole system around the crossbar constant.

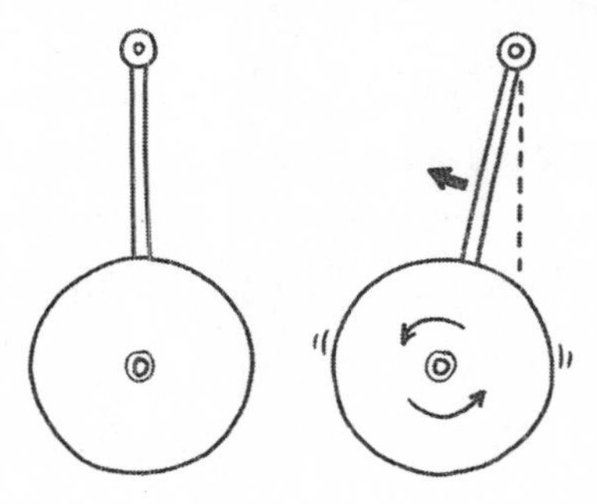

Pumping a swing works the same way. When you twist your body while holding onto the chains, the swing twists a little in the opposite direction, pushing you up against gravity. Then, once gravity reverses your direction, you twist your body back the other way. Since you're moving in the other direction, the twist pushes you a little more in your direction of motion. By twisting at the right part of your swing, both the forward and backward twists make you swing a little higher.

If the swing set is really tall, pumping becomes less efficient. When you're really far away from the crossbar, your rotation doesn't impart much of a twist on the whole system and the swing moves less in response. An adult leaning back once on an 8-foot swing might rotate the swing around the pivot by 1 degree, but the same motion on a 30-foot swing would nudge it by just 0.07 degrees.

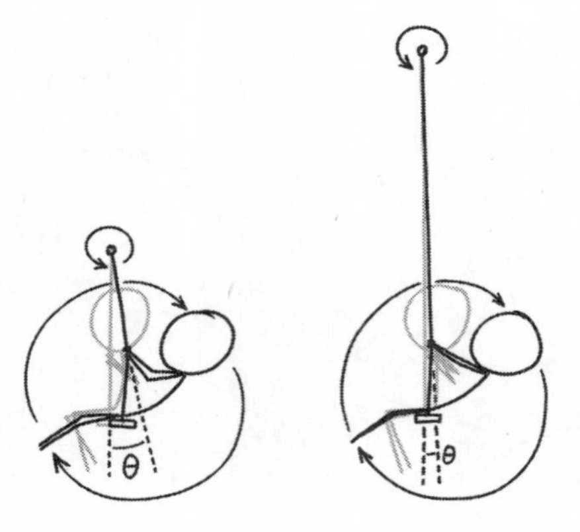

The reduced efficiency of pumping on a taller swing set means it takes longer to get the swing in motion. On an 8-foot swing, each pump adds a little over a degree, so if you want to get up to a good 45-degree swing, it will only take 45 pumps, which will take about 70 seconds. But on a 30-foot swing, where each pump adds so much less to your arc, you'll need 640 pumps to get up to 45 degrees. Since a taller pendulum takes longer to swing back and forth, you'll kick fewer times per minute, so those 640 pumps will take over half an hour.

If you try this on a real 30-foot swing, you'll find that you can't get up to 45 degrees at all. In fact, you won't be able to get as high off the ground as you could on an 8-foot swing! Thanks to air resistance, you lose a little speed while you're in the bottom part of each swing. When you take larger swings, you go faster, and experience more drag in the middle. When you swing up to about 20 degrees, you lose more energy to drag than you gain by pumping. An 8-foot swing can actually carry you higher than a 30-foot one!

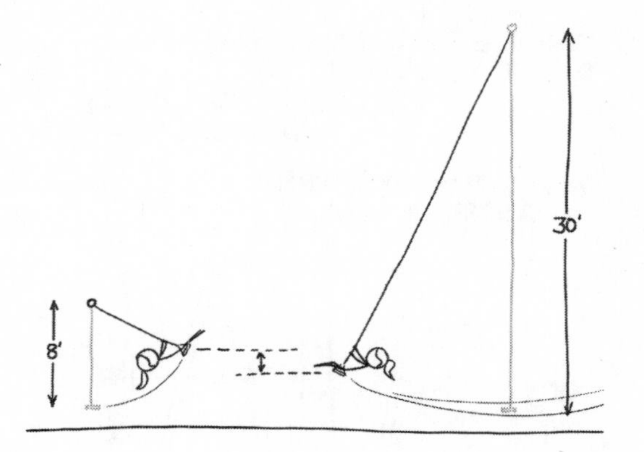

There are some very big swings out there. At the Moses Mabhida Stadium in Durban, South Africa, visitors can climb out on a walkway high above the field and take a swing on a 200-foot rope dangling from the scaffolding over the stadium. But at those speeds, air resistance takes its toll—when riders reach the bottom, they've lost most of their momentum, so they don't swing very far back up the other side. Kicking their feet won't help; the swing is so tall that pumping has virtually no effect.

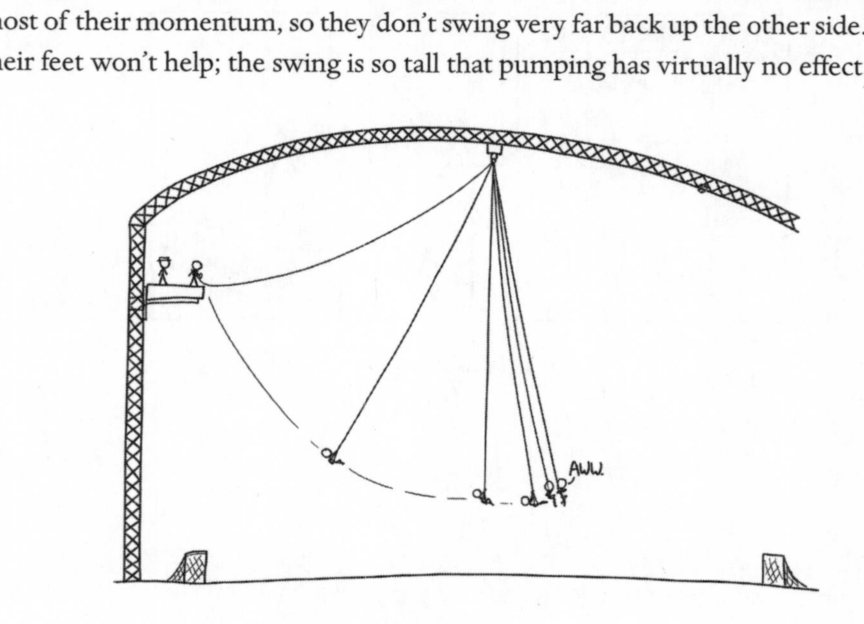

A giant swing might be fun, but it won't help you get closer to space. Using measurements for an average rider, the ideal swing for reaching maximum altitude turns out to be 10 or 15 feet—exactly the size of a large playground swing.

Once again, the kids have things figured out.

18. AIRLINER CATAPULT

My friend is a commercial airline pilot. She says that a significant amount of fuel is spent on takeoff. To save fuel, why couldn't we launch airplanes using catapult systems like on aircraft carriers (calibrated to normal human accelerations)? Could significant amounts of fossil fuels be saved if the catapults could be run by some other clean energy? I'm imagining a rope . . . one end tied to the airplane, the other tied to a large boulder at the edge of a cliff. Just push the boulder off the cliff!

—Brady Barkey, Seattle, WA

I like how this question starts off sounding very cool and futuristic, and ends with boulders and strings.

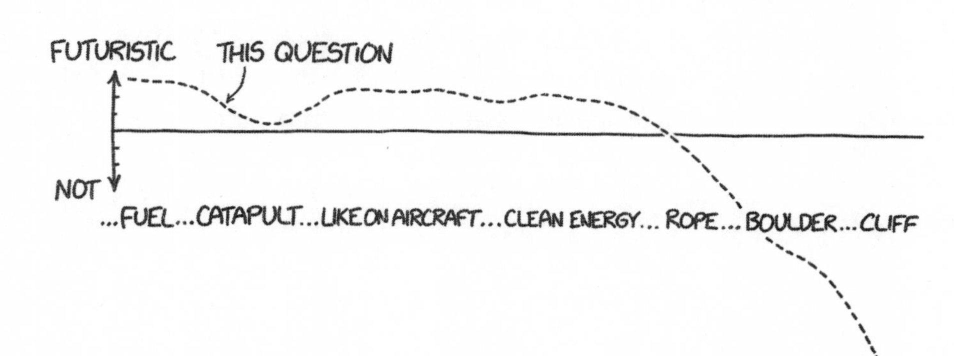

It's true that airliners burn fuel faster while they're taking off, but takeoff is brief. A small airliner like the Airbus A320 might burn just 10 or 20 gallons of fuel while accelerating down the runway to takeoff speed compared to thousands of gallons during the rest of the flight.

The plane continues to burn fuel quickly during the climb to cruising altitude, which takes quite a bit longer than the acceleration down the runway. This fuel can add up; for an A320 it might be several hundred gallons. But a catapult can only help you while you're on the ground. If it could keep helping you during the climb, it wouldn't be a catapult, it would be an escalator.

You could use the catapult to gain extra speed while still on the ground. When an airliner takes off, it's usually traveling at less than half of its cruising speed. Using a catapult to gain more speed near the ground would mean burning less fuel to get up to speed during the climb.

There are two problems with this.* The first is that drag from the dense air near the surface will make you lose some of that speed before you can get up into the upper atmosphere.

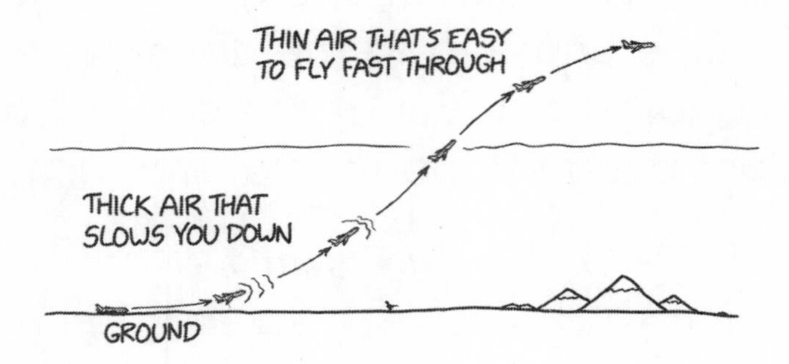

The second, bigger problem is real estate.

Airliners generally accelerate forward at about 0.2g or 0.3g during takeoff, which is why they typically need a runway that's at least a mile long to take off. If you're willing to go all the way up to 0.5g, similar to the acceleration you'd feel in a fast car with the gas pedal pushed all the way down, you could in theory get away with barely half a mile. But if you want to accelerate up to near full cruising speed before taking

---

* I mean, at least.

off, giving you enough momentum to coast up out of the thickest part of the atmosphere, you'll need a runway that's *nine times longer*. Even if we don't leave any safety margin, that means a runway that's at least 4½ miles long.

Here's what the Washington, DC, airport would look like if you extended its main runway to that length:

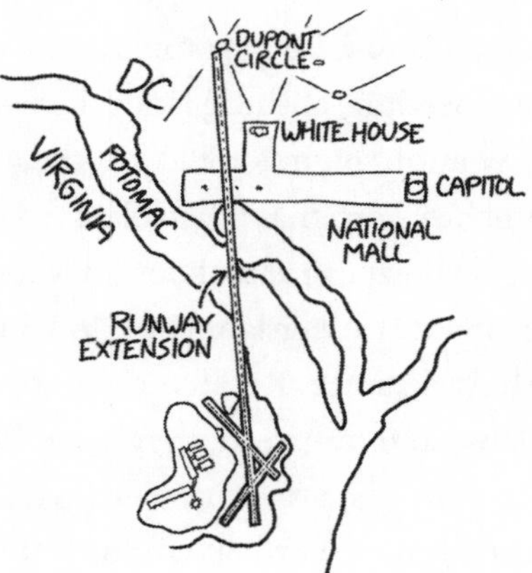

The runway would cross the National Mall between the Lincoln Memorial and the Washington Monument—just missing the FDR Memorial and the World War II Memorial—and then continue through the city, ending somewhere near Dupont Circle.

To be fair, the idea of a catapult launch for passenger planes isn't totally ridiculous. The fuel savings may be small, but they could allow bigger planes to take off on shorter runways. They could also make launches quieter; noise is a perennial problem for urban airports.

There have been a few serious airliner catapult proposals. In 1937, NACA—the predecessor to NASA—studied land-based catapult launches to help gigantic passenger planes take off without needing absurdly long runways.* In 2012, Airbus published concept art for what aviation might look like in the year 2050. Their concept art included a catapult-like launch system that they called Eco-climb.

But outside of the occasional experimental design, catapults have been limited to specialized situations, like aircraft carrier launches, in which planes need to accelerate fast to take off over a short distance. Because the potential fuel savings are pretty small compared to the expense and overhead, they seem likely to stay that way.

If you insist on building your system—complete with the rope and cliff—here's a tip: To accelerate a 200-ton airliner up to 400 mph, you'd need an extremely heavy counterweight or an extremely tall cliff. A huge thousand-ton weight would need to fall the height of a supertall skyscraper.

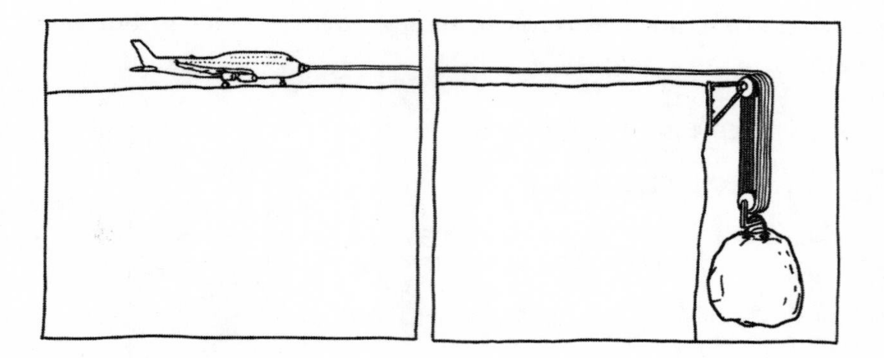

* Of course, to the folks in 1937, "gigantic" planes held 40 people, and the "absurdly long" runways they imagined were less than a mile long—nothing compared to the several-mile-long runways we ended up building.

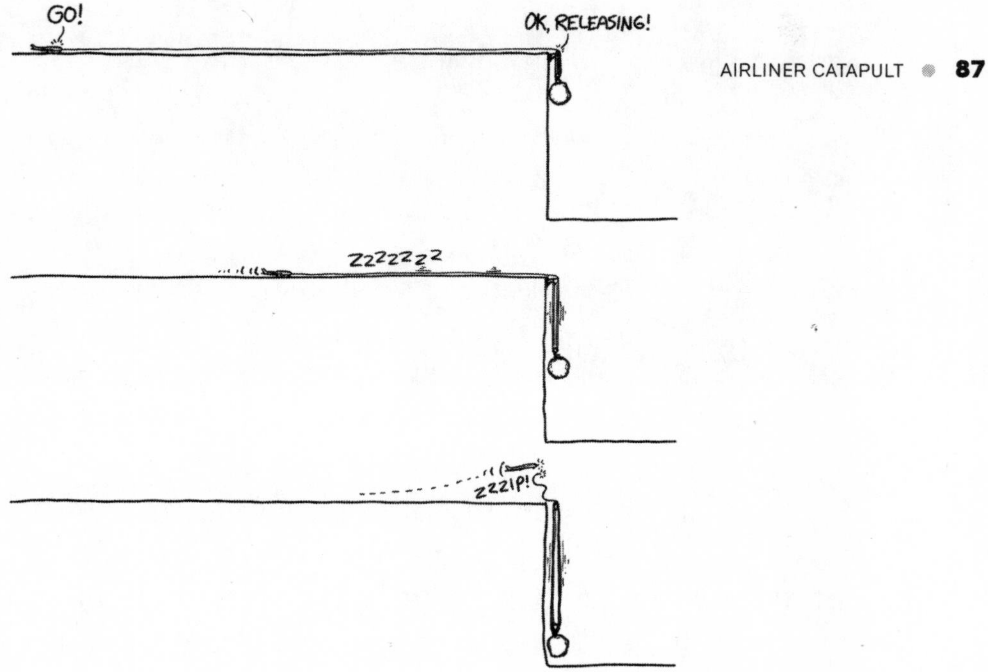

If you used a heavier weight, you wouldn't need to drop it as far. Now, I'm not suggesting anything specific here, but for the record, the aboveground portion of the Washington Monument weighs about 80,000 tons. An 80,000-ton object would only need to drop a short distance to accelerate an airliner up to takeoff speed.

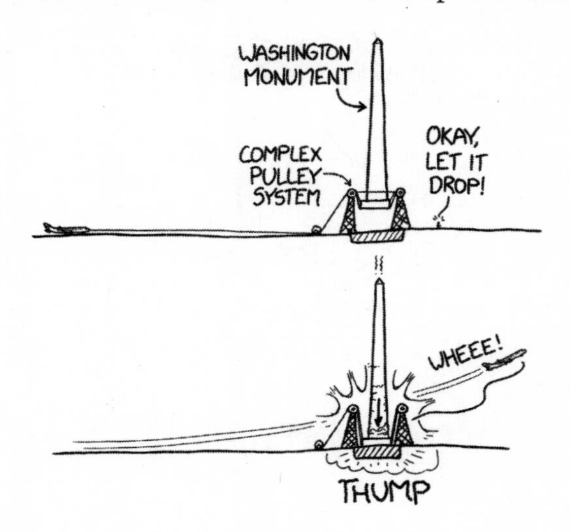

Just a thought.

short answers #2

Q Billy the Clown is running out of cash, so in order to raise money, he devises his newest trick: He will inflate, by mouth, a standard-size party balloon until the material (some form of indestructible rubber) is just one atom thick. How large would the inflated party balloon be?

—Alan Fong

IT'S A TOTAL MYSTERY WHY BILLY IS RUNNING OUT OF CASH.

Q How many leaf blowers would it take to move a standard SUV?

—Ashley H.

On level ground, with the vehicle in neutral, you could probably get it moving with just one or two dozen heavy-duty leaf blowers, although you'd need a lot more if you wanted to accelerate it fast enough to avoid getting honked at.

Q If you put a vacuum at extremely high suction power and aimed it at a normal BMW sedan, what would happen?

—Anonymous

Q On a warm summer evening, when you sit outside with a light on, you can be quite sure that bugs will be attracted to the light. Then why is it that these same bugs don't fly toward the biggest and strongest lamp of them all, namely the Sun, during the day?

—Anonymous

The question of why moths and other insects fly toward lamps is something of an open question in entomology, but the question of why they *don't* all fly toward the Sun has a much simpler answer:

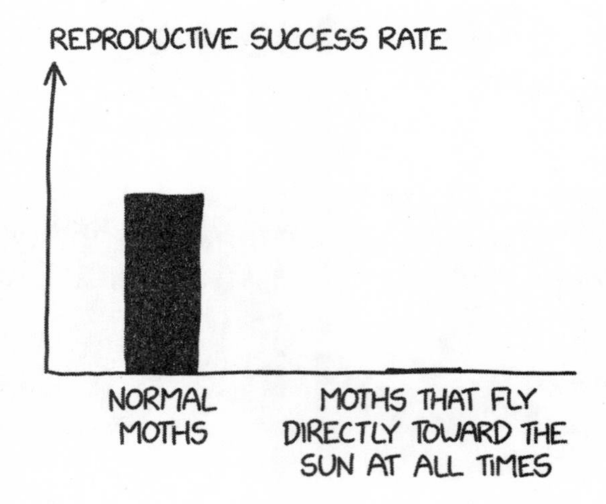

Q If you collected all the guns in the world and put them on one side of the Earth, then shot them all simultaneously, would it move the Earth?

—Nathan

No, although in my personal opinion, if you could get them to stay there, it would make the other side of the Earth a nicer place to live.

Q What would happen if you microwaved a smaller microwave, while the smaller one was on as well?

—Michael

You would no longer be welcome in that IKEA.

Q If you're jumping on a trampoline, how fast would your body have to be going to:
A. Break all bones on impact
B. Make your body go through the tiny holes of the mesh.
—Micah Lane

A: Breaking *all* the bones in your body is hard, because a lot of them are pebble-size and embedded deep inside larger body structures. I don't know exactly how fast you'd have to go to break all of them, but it would definitely be fast enough that the trampoline wouldn't make much of a difference.

B: I am happy to report that this can't happen.

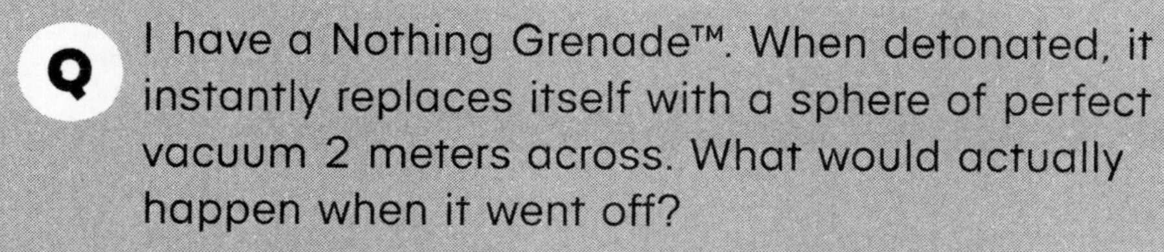

Q I have a Nothing Grenade™. When detonated, it instantly replaces itself with a sphere of perfect vacuum 2 meters across. What would actually happen when it went off?

—Dave H.

The sphere of vacuum would collapse, colliding in the center with so much force that it would rapidly heat and might even turn briefly to plasma. Energy would radiate outward both in the form of a pulse of heat and a shock wave, capable of causing severe injury or death and of destroying small structures.

In other words, what you have is a regular grenade.

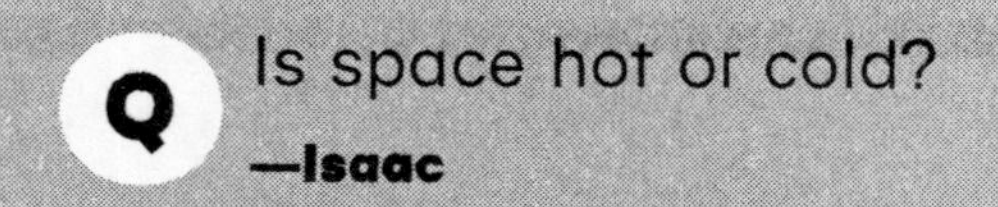

Q Is space hot or cold?

—Isaac

Based on the textbook definition of temperature, space is hot, at least here in the Solar System. The molecules in space are individually moving very fast, which means that each one has a lot of energy, and temperature is usually defined as the average kinetic energy of the molecules in a substance. But there are so few molecules in space that even though each one has a lot of energy, the total amount of heat energy is small, which means it doesn't warm things up very much. It may be warm in theory, but it feels cold in practice.

Space may be hot, but it's the hottest place you can freeze to death.

> **Q** How many bones can you remove from the human body while allowing the human to continue living? Asking for a friend.
>
> **—Chris Rakeman**

I don't think this person is really your friend.

> **Q** What would happen if you put a human under a g-force of 417 Gs for twenty seconds?
>
> **—Nythil**

You would be arrested for murder.

Where or how can one commit a murder and not be prosecuted for it?

—Kunal Dhawan

There's a famous legal article by law professor Brian C. Kalt arguing that there is a 50-square-mile area of Yellowstone National Park in which people can commit felonies with impunity. The Constitution has clear rules about where juries must come from, but because of a mistake in drawing district lines, prosecuting a crime in this area requires that the jury come from an area with a population of 0.

But don't head out on a crime spree just yet. I asked a federal prosecutor about the "Yellowstone loophole." He laughed, then said that you would absolutely be prosecuted if you tried to take advantage of it. I brought up the arguments that Professor Kalt made in his article. He replied, and I quote, "Law professors say a lot of stuff."

I read today that insects make at least $57 billion a year for the US economy. If we were to pay every single insect in the United States equally for their economic contribution, how much would each insect get?

—Hannah McDonald

Estimates of economic value are complicated and depend a lot on definitions, but for the sake of the question we'll take that $57 billion at face value. Some of those insects are probably pulling a lot more weight than others—personally, I feel that ants do an awful lot of work—but let's assume we're going to pay every insect equally.

How many insects are there? In the 1990s, Jan Weaver and Sarah Heyman of the University of Missouri conducted a survey that found about 2,500 insects per square meter of Missouri's Ozark forests. Other surveys have found higher numbers, either because they looked in different types of forest, dug deeper into the soil, or were able to count smaller insects. But the surveys are generally in relatively rich areas, and the national average might also be a lot lower than the average for leaf litter on a forest floor. If we just take their figure as a loose estimate of the national average, that implies there are about 20 quadrillion individual insects in the United States.

If we divide that $57 billion up among 20 quadrillion insects, each one will receive $0.0000029, or one penny per 3,500 insects. Coincidentally, the average weight of an insect in the survey was a little under 1 mg, so those 3,500 insects would weigh about as much as the penny they would receive.

Based on the prevalence in the Weaver and Heyman survey, the money would be divided up as follows:

- **$18 billion** to flies, including mosquitoes
- **$16 billion** to bees, wasps, and ants
- **$10 billion** to beetles
- **$7 billion** to thrips, tiny insects that drink fluid from plants
- **$1 billion** to butterflies and moths
- **$1 billion** to the true bugs
- **$4 billion** divided among the rest

Looks good to me! But for the record, if I'm ever put in charge of this budget, the first thing I'm going to do is cut the funding to mosquitoes.

Q What, in today's world and yesterday's world, does it mean to be human, in all social and biological factors?

—Seth Carrol

I think you meant to submit this to *Why If?*.

why if?
DEEPLY UNGRAMMATICAL ANSWERS
to Unanswerable Philosophical Questions

What if an object like the Chicxulub impactor hit Earth with a relatively low relative speed of (let's say) 3 mph?

—Beni von Alemann

It wouldn't cause a mass extinction, but that would be a small consolation for anyone who was standing near it when it landed.

Sixty-six million years ago* a big rock from space hit the Earth near the present-day city of Mérida, Mexico. This impact led to the extinction of most of the dinosaurs.

Anything from space that hits the Earth is going fast by the time it reaches the ground. Even if an object is drifting along slowly when it encounters the Earth, the fall down into the planet's gravity well will accelerate it up to at least escape velocity. That speed gives objects a lot of kinetic energy, which is why pebble-size meteors burn so brightly and why larger rocks can punch big holes in the crust.

A slow meteor would be different. Let's say you carefully lowered a meteor down until it was hovering just 5 inches above the surface, then let go.

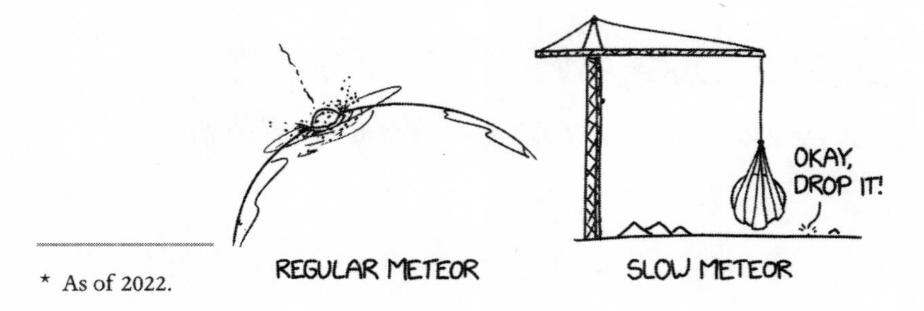

* As of 2022.

The meteor would start to fall, just like any object. After a tenth of a second, it would make contact with the ground.

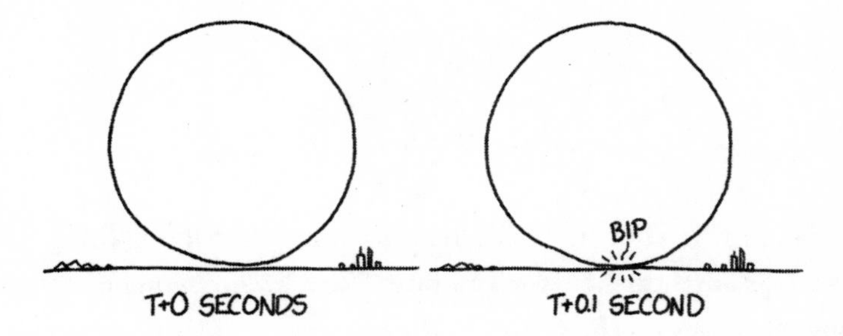

When the bottom of the meteor touched the ground, it would be traveling at about 3 miles per hour, less than a thousandth of the speed of the real dinosaur-killing meteor. The bottom part of the meteor might come to a stop against the ground, but the 10 kilometers above it would keep falling.

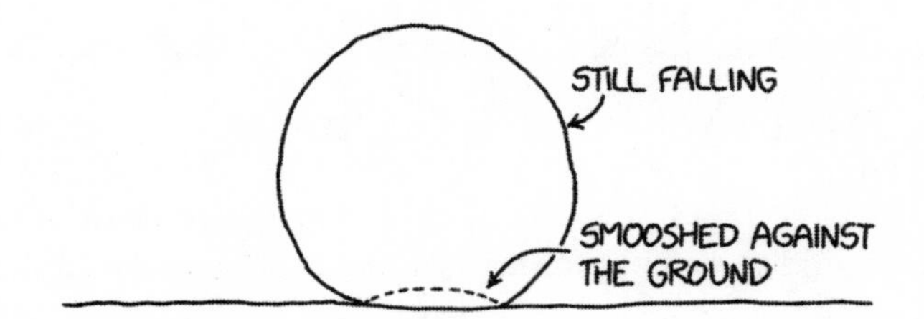

Most comets and asteroids aren't very strong. We used to imagine asteroids as potato-shaped solid rocks pockmarked with craters. It's true that some asteroids look like that, but now that we've visited several of these objects with robotic probes, we know that a lot of them are more like heaps of gravel loosely held together by gravity and frost. They're more like sandcastles than boulders.

If you google "world's biggest ball of sand," you won't find much,* because it's hard to make a ball of sand bigger than a softball. Even if you try to use sand with just the right amount of water and pack it very carefully, you'll find that a larger ball

* This was true when I wrote it, but will probably change by the time you read it. If you found this book by googling "world's biggest ball of sand" and couldn't figure out why it was at the top of the results, well, you've finally solved the mystery!

of sand can't support its own weight. The same thing that happens to a ball of sand would happen to the impactor.

"Soil liquefaction" is a boring-sounding phrase for a terrifying thing. Under certain conditions, such as earthquakes, soil can flow like a liquid, which is extremely alarming for anyone who lives on the ground. The material in the impactor would undergo this same transformation, flowing outward across the surface in an omnidirectional landslide of supersonic soil liquefaction.*

Over the next 45 seconds, the meteor would go from a falling ball to a spreading disk.

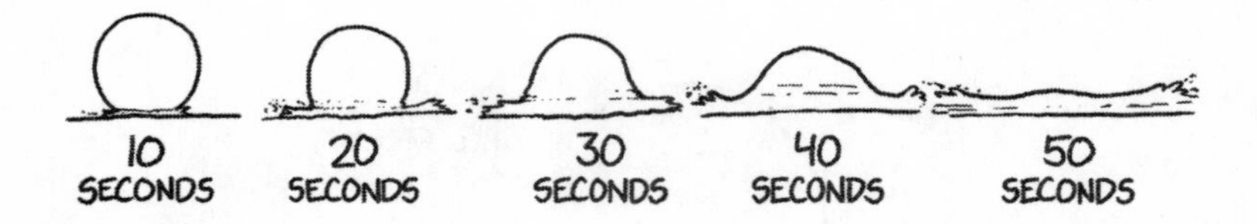

The landslide would spread out for miles. Studies of large landslides on Earth and on other bodies in the Solar System show that the area covered by a landslide depends mainly on the total original volume of material, and not the exact details of how it's deposited. This tells us that our landslide would spread out about 30 or 40 miles from the original contact point—perhaps a bit more, since it would have a higher speed than most landslides. If it happened in the same place as the Chicxulub impact, it would probably cover most of the area of the original crater.

---

* I searched several research paper archives for "supersonic soil liquefaction" and was disappointed to find no results. Maybe someone out there is working on a grant proposal.

The Chicxulub impact site is along the coast, so much of the debris from our meteor would flow into the ocean. Just like the original impact 66 million years ago, this one would displace a huge amount of seawater.

The Cretaceous impact kicked up a tsunami that swept across the Gulf of Mexico and traveled many miles inland. The impact also shook the Earth hard enough to make water slosh around across the planet, creating tsunami-like waves in lakes that weren't even connected to the Gulf of Mexico.

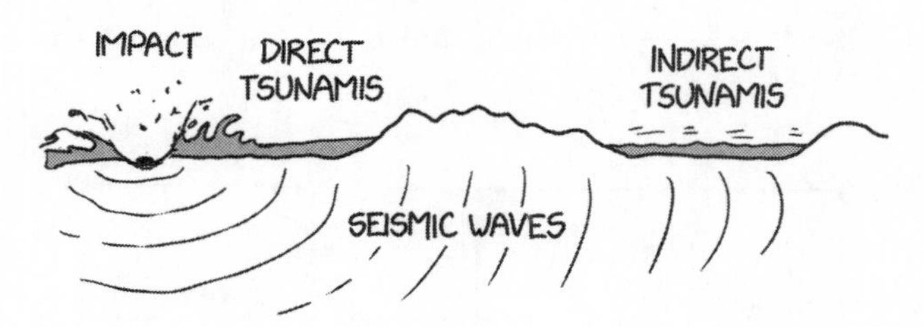

The shaking from our impact wouldn't be as severe as the Cretaceous one, because our impactor would be so much slower. Our impact would be equivalent to a magnitude 7 earthquake, compared to the magnitude 10+ of the one in the Cretaceous, and our tsunami would be smaller as well.

However, don't rush to watch from the Gulf coast; the wave might not be *that* much smaller. Most of the energy of the Cretaceous impactor went into creating the crater, and a relatively small fraction went into creating a tsunami. But pouring a lot of material into the ocean—rather than vaporizing a hole in the water and letting it fill in—can be a more efficient way to generate waves, so our tsunami might reach quite some distance inland.

The landslide itself would bury the city of Mérida. Half an hour later, the tsunami would destroy the rest of the cities bordering on the Gulf of Mexico. Over the next few hours, smaller waves would ripple around the world ocean before gradually subsiding.

If you were living on the other side of the world—say, in Jakarta or Perth—and you were away from the shore during the brief coastal flooding, you wouldn't notice much else. Unlike 66 million years ago, there wouldn't be global firestorms from ejected debris reentering the atmosphere. No volcanic eruptions would be triggered. There would be some dust thrown into the air, but there wouldn't be global cooling from volcanic aerosols.

The slow impact wouldn't cause a mass extinction, but it could still cause an extinction.

Isla Nublar, the fictional location of Jurassic Park, is located off the southwestern coast of Costa Rica. The size of the island isn't established in the original film, but John Hammond mentions that he's installed "50 miles of perimeter fence," which means the park has an area of less than 200 square miles.

If humans really did clone dinosaurs, and if the location of the impact shifted about a thousand miles to the south . . .

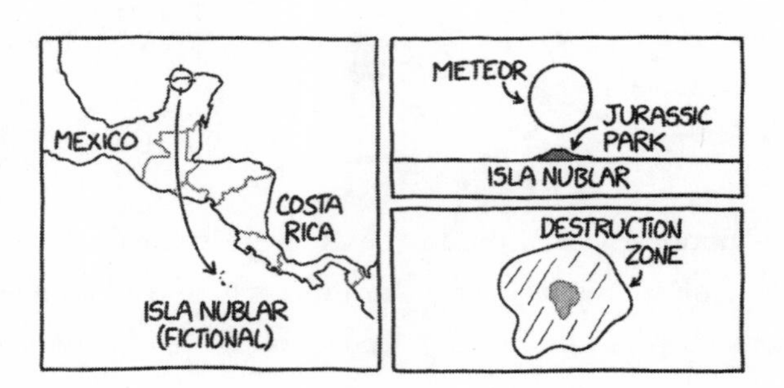

. . . it *could* cause a dinosaur extinction.

20. ELEMENTAL WORLDS

What if Mercury (the planet) were entirely made of mercury (the element)? What if Ceres was made of cerium? Uranus made of uranium? Neptune made of neptunium? What about Pluto made of plutonium?

—Anonymous

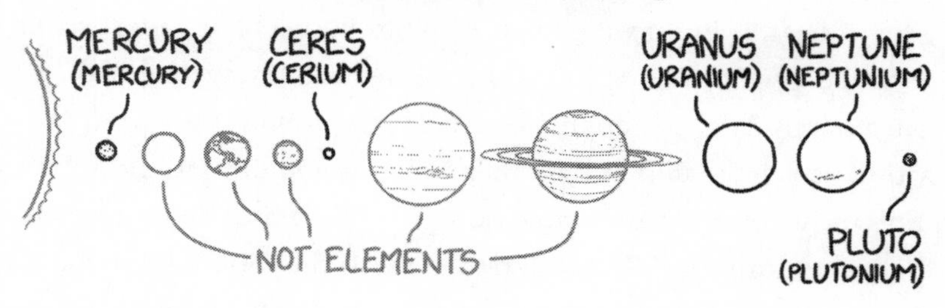

There are five large worlds that share names with elements: the planets Mercury, Uranus, and Neptune, and the dwarf planets Ceres and Pluto.

From our point of view here on Earth, Mercury and Ceres wouldn't change that much. Mercury would get more than twice as heavy and about five times brighter, thanks to its shiny new semiliquid surface. Ceres would get three times heavier and almost ten times brighter—bright enough to see with the naked eye under dark skies.

Unfortunately, dark skies would get a little harder to find, thanks to the other three planets.

The changes to the other three element-named worlds—Uranus, Neptune, and Pluto—would be a little more dramatic.

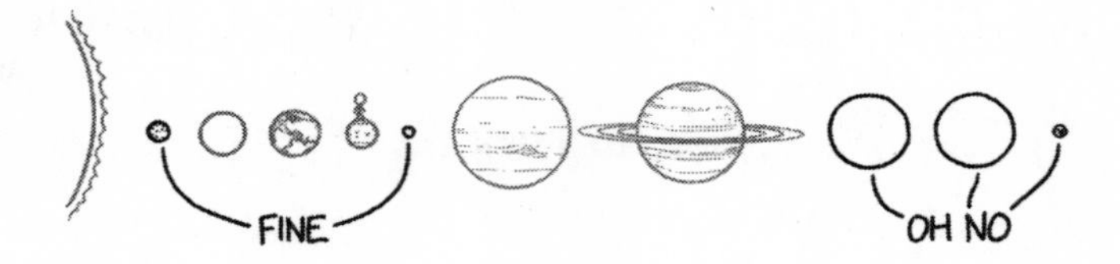

Uranium, plutonium, and neptunium are all radioactive, so these planets would produce a lot of heat. If Pluto were made of plutonium-244, the most stable isotope, its surface would get hot enough to glow the reddish-orange color of a campfire, making it just barely bright enough to see from Earth with the naked eye—though only a few times a year, thanks to the other two new additions to the Solar System.

Uranium's most common and stable isotope is ^{238}U, which decays very slowly over billions of years. A lump of ^{238}U wouldn't be hot to the touch—you could handle it without any risk of radiation poisoning. But if you collected it into a planet-size ball, the tiny amount of heat produced by each part would add up to heat the planet to thousands of degrees.*

It might seem strange that a metal that's cool to the touch in small amounts would be so hot when collected together in a big ball, but this is just a consequence of scale. Since volume grows faster than surface area, larger objects produce more heat per unit of surface area, so they have to get hotter to radiate it away. Really big objects can get extremely hot from even a tiny amount of heat production per unit of volume.

* Fahrenheit, Celsius, Kelvin, it's true in any of them.

Even the core of the Sun, where nuclear fusion happens, would be pretty cold if you could somehow isolate a piece of it. A cup of solar core material* produces about 60 milliwatts of thermal energy. By volume, that's about the same heat production rate as the body of a lizard, and less than that of a human. In a sense, you're hotter than the Sun—there's just not as much of you.†

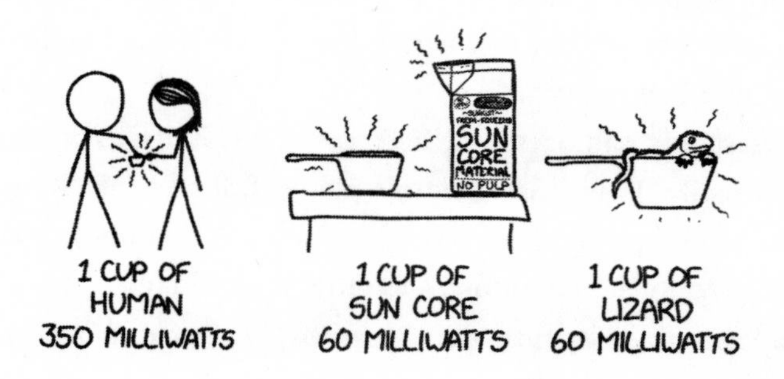

The real Uranus, lit by reflected sunlight, is too dim to see with the naked eye, although you can spot it with binoculars if you're lucky. The superhot uranium Uranus would glow brightly, and would be visible in the sky like an ordinary star.

Neptune would be the real problem.

	BEFORE	AFTER
MERCURY	VISIBLE	VISIBLE
CERES	NOT VISIBLE	VISIBLE
URANUS	BARELY VISIBLE	VISIBLE
NEPTUNE	NOT VISIBLE	*OW, MY EYES!*
PLUTO	NOT VISIBLE	VISIBLE

Neptunium is not something you run into every day. Uranium and plutonium aren't exactly *common*, but they're well-known enough thanks to their role in nuclear weapons. Neptunium—one of their neighbors in the periodic table—is significantly more obscure.

---

* If you find a recipe that calls for this, do not make it.

† Unless you're a lizard, in which case, hi, thank you for crawling onto this book! I hope this page was left open in the sun so it's nice and warm.

It does pop up now and then. In early 2019, a middle school in southern Ohio abruptly closed in the middle of the school year. The reason? Neptunium contamination. The school is located a few miles from the Portsmouth Gaseous Diffusion Plant, a former processing site for nuclear fuel that stopped operating in 2001. In early 2019 the district was notified that Department of Energy air monitors across the street from the school had picked up excess neptunium, a possible by-product of waste disposal at the plant. The district shut down the school immediately, and it remained closed through the following year.*

Neptunium is highly radioactive. Microscopic amounts of it can be dangerous enough, but you really don't want a whole planet full of it. If Neptune were made of neptunium, it would produce *far* more heat than its neighbors Uranus and Pluto. Neptune would not only get hot enough to glow, it would produce so much heat that it would actually vaporize, forming a thick atmosphere of gaseous neptunium.

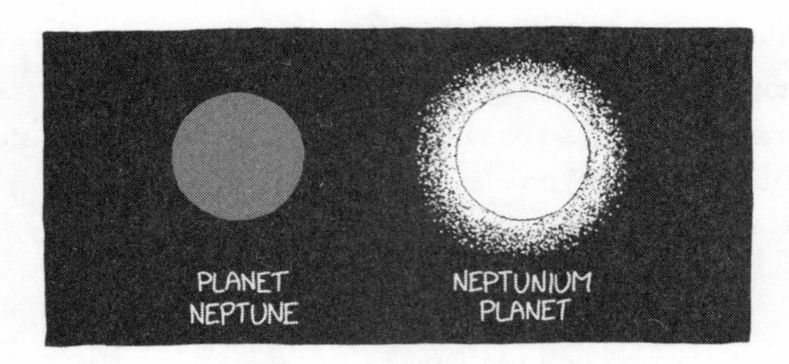

Neptune would be as bright as a medium-size star. It wouldn't quite outshine the Sun—which is on the bright side, as stars go—but Neptune's surface would be hotter than the Sun's, so its color would be bluer.

Neptune is much farther from us than the Sun, so its apparent brightness would be reduced, but it would still be about as luminous as the full moon.

WHEN NEPTUNE HITS YOUR EYE LIKE A BIG PIZZA PIE
THAT'S AN X-RAY

Unlike the Moon, Neptune wouldn't go through monthly cycles. Since it takes over a century and a half to orbit the Sun, it would appear in roughly the same

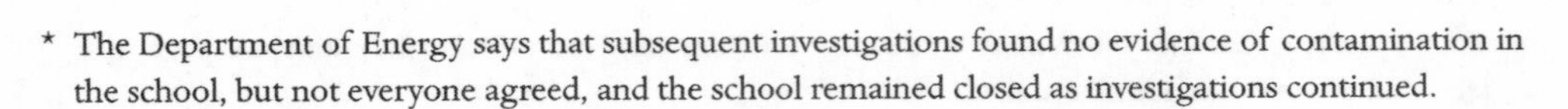

* The Department of Energy says that subsequent investigations found no evidence of contamination in the school, but not everyone agreed, and the school remained closed as investigations continued.

place among the stars every night for years on end. In the 2020s, Neptune would be in the sky for most of the night from June through December, washing out the constellations Aquarius, Pisces, and Pegasus. Over the next few decades, it would move lazily across the sky through Aries and Taurus. Its light would render Orion virtually invisible for several decades.

Other than some astronomical and astrological complications, life on Earth would probably continue without too much trouble. The interior of the newly radioactive planets would get hot, but none of them would get hot enough to cause catastrophic energy release through fission. And our atmosphere would protect us from whatever exotic particles came flowing toward Earth from the direction of Pisces.

The situation would be extremely different if we didn't use stable isotopes. If Uranus were made of ^{235}U instead of ^{238}U, it would be a lot worse. Any lump of ^{235}U larger than a softball is big enough to undergo a fission. ^{235}Uranus would instantly undergo a runaway chain reaction, converting the whole planet into an expanding cloud of high-energy particles and X-rays. A little under three hours later, the shock wave would reach—and completely obliterate—the Earth, stripping away its surface and leaving it a molten blob hanging in the sky.

There's a lesson here: If you have a choice between isotopes and you're not sure which to pick, go for the most stable one.

21. ONE-SECOND DAY

What would happen if the Earth's rotation were sped up until a day only lasted one second?

—Dylan

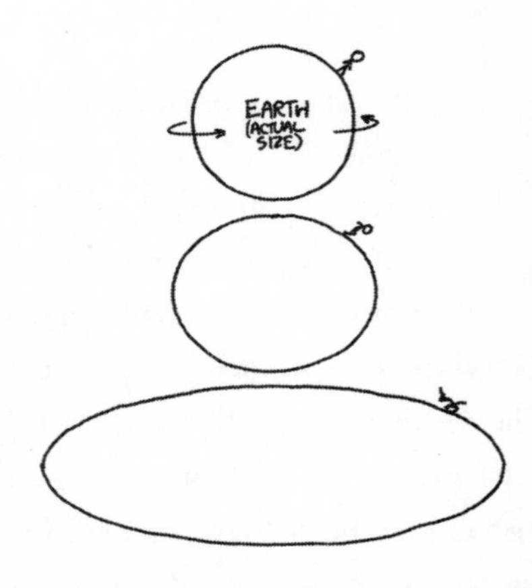

That would be apocalyptic, but there would be a brief period every two weeks when it would be even *more* apocalyptic.

The Earth rotates,*[citation needed]* which means its midsection is being flung outward by centrifugal force. This centrifugal force isn't strong enough to overcome gravity and tear the Earth apart, but it's enough to flatten the Earth slightly and make it so you weigh almost a pound less at the equator than you do at the poles.*

If the Earth (and everything on it) were suddenly sped up so that a day only lasted one second, the Earth wouldn't even last a single day.† The equator would be moving at over 10 percent of the speed of light. Centrifugal force would become much stronger than gravity, and the material that makes up the Earth would be flung outward.

* This is due to a combination of several effects, including centrifugal force, the flattened shape of the Earth, and the fact that if you go far enough toward the pole in North America people start offering you poutine.

† Either kind.

You wouldn't die instantly—you might survive for a few milliseconds or even seconds. That might not seem like much, but compared to the speed at which you'd die in other *What If* scenarios involving relativistic speeds, it's pretty long.

The Earth's crust and mantle would break apart into building-size chunks. By the time a second* had passed, the atmosphere would have spread out too thin to breathe—although even at the relatively stationary poles, you probably wouldn't survive long enough to asphyxiate.

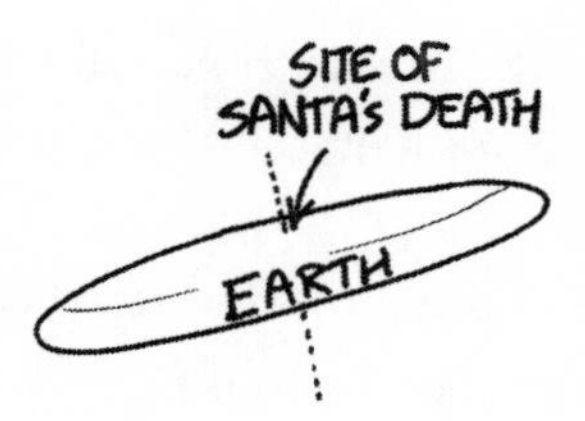

In the first few seconds, the expansion would shatter the crust into spinning fragments and kill just about everyone on the planet, but that's relatively peaceful compared to what would happen next.

Everything would be moving at relativistic speeds, but each piece of the crust would be moving at close to the same speed as its neighbors, so there wouldn't be any immediate relativistic collisions. This means things would be relatively calm . . . until the disk hit something.

The first obstacle would be the belt of satellites around the Earth. After 40 milliseconds, the International Space Station (ISS) would be struck by the edge of the expanding atmosphere and instantly vaporized. More satellites would follow. After a second and a half, the disk would reach the belt of geostationary satellites orbiting above the equator. Each one would release a violent burst of gamma rays as the Earth consumed it.

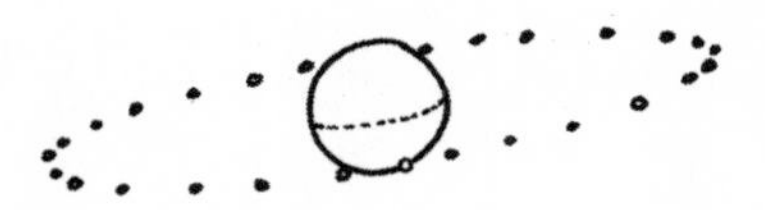

* I mean, a day.

The debris from the Earth would slice outward like an expanding buzz saw. The disk would take about 10 seconds to pass the Moon, another hour to spread past the Sun, and would span the Solar System within a day or two. Each time the disk engulfed an asteroid, it would spray a flood of energy in all directions, eventually sterilizing every surface in the Solar System.

Since the Earth is tilted, the Sun and the planets aren't usually lined up with the plane of the Earth's equator, so they'd have a good chance of avoiding the terrestrial buzz saw.

However, every two weeks, the Moon crosses the plane of the Earth's equator. If Dylan sped up the Earth at this moment, the Moon would be right in the path of the expanding disk.

The impact would turn the Moon into a comet, rocketing out of the Solar System on a wave of high-energy debris. The flash of light and heat would be so bright that if you were standing on the surface of the Sun, it would be brighter above you than below. Every surface in the Solar System—Europa's ice, Saturn's rings, and Mercury's rocky crust—would be scoured clean . . .

. . . by moonlight.

22. BILLION-STORY BUILDING

My daughter—age 4½—maintains she wants a billion-story building. It turns out not only is it hard to help her appreciate this size, I am not at all able to explain all of the other difficulties you'd have to overcome.

—Keira, via Steve Brodovicz, Media, PA

Keira,

If you make a building too big, the top part is heavy and it squishes the bottom part.

Have you ever tried to make a tower of peanut butter? It's easy to make a little tiny one, like a blobby castle on a cracker. It will be strong enough to stay up. But if you try to build a really big castle, the whole thing smushes flat like a pancake.

Note to Keira: If your dad tells you not to build stuff with peanut butter, don't listen to him. If he complains about the mess on the table, then just sneak jars into your room and build the tower on the carpet there. You have my permission.

The same thing that happens with peanut butter happens with buildings. The buildings we make are strong, but we couldn't make one that went all the way up to space, or the top part would squish the bottom part.

We can make buildings pretty tall. The tallest buildings are almost 1 kilometer tall, and we could probably make buildings 2 or even 3 kilometers tall if we wanted, and they would still be able to stand up under their own weight. Higher than that might be tricky.

But there would be other problems with a tall building besides weight.

One issue would be the wind. The wind up high is very strong, and buildings have to be very strong to stand up against the wind.

Another big problem would be, surprisingly, elevators. Tall buildings need elevators, since no one wants to climb hundreds of flights of stairs. If your building has lots of floors, you need lots of different elevators, since there would be so many people trying to come and go at the same time. If you make a building too tall, the whole thing gets taken up by elevators and there's no space for regular rooms.

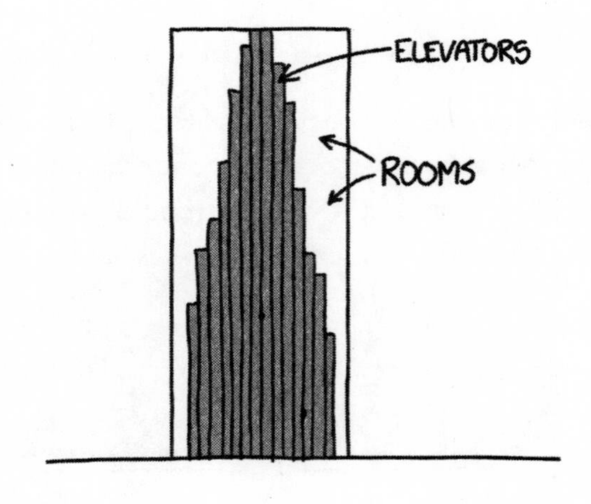

Maybe you can think of a way to get people to their floors without having too many elevators. You could try pigeons, like in chapter 6. You could make a giant elevator that takes up 10 floors. You could make fast elevators that travel like roller coasters. You could fly people up to their rooms with hot-air balloons. Or you could launch them with catapults.

Elevators and the wind are big problems, but the biggest problem would be money.

To make a building really tall, someone has to spend a lot of money, and no one wants a really tall building enough to pay for it. A building many miles tall would cost billions of dollars. A billion dollars is a lot of money! If you had a billion dollars, you could buy a spaceship, save all the world's endangered lemurs, give a dollar to everyone in the United States, and still have some left over. Most people don't think giant towers a few miles tall are important enough to spend a lot of money on.

If you got really rich, so you could pay for a tower to space yourself, and solved all the engineering problems, you'd still have problems making a tower a *billion* stories tall. A billion stories is just too many.

A big skyscraper might have about 100 floors, which means it's as tall as 100 little houses.

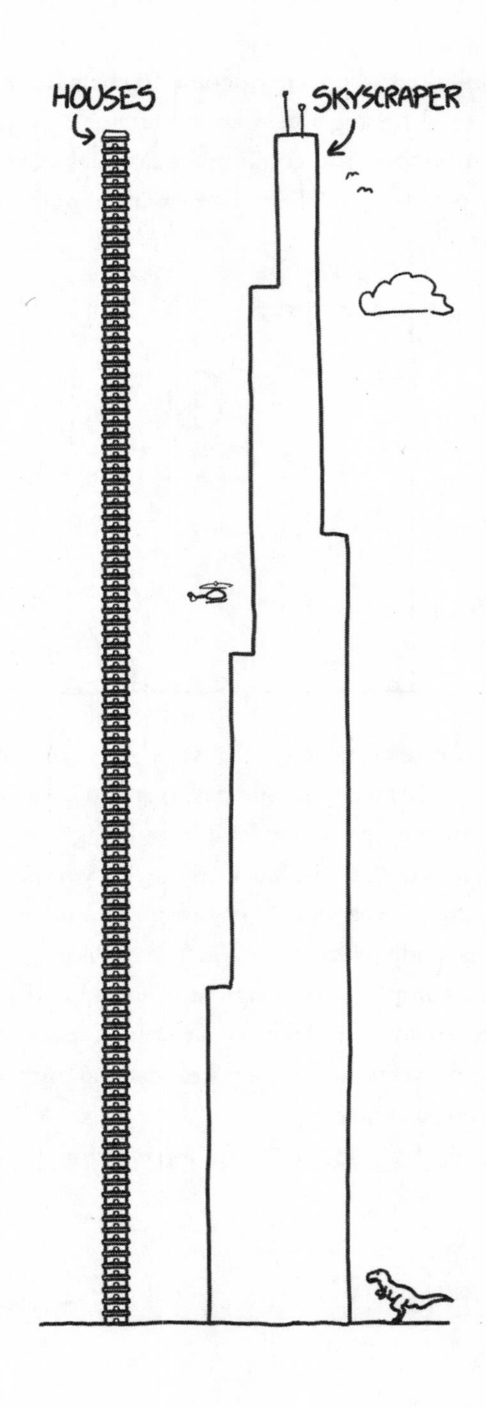
HOUSES
SKYSCRAPER

If you stacked 100 skyscrapers on top of one another to make a mega-skyscraper, it would reach halfway to space:

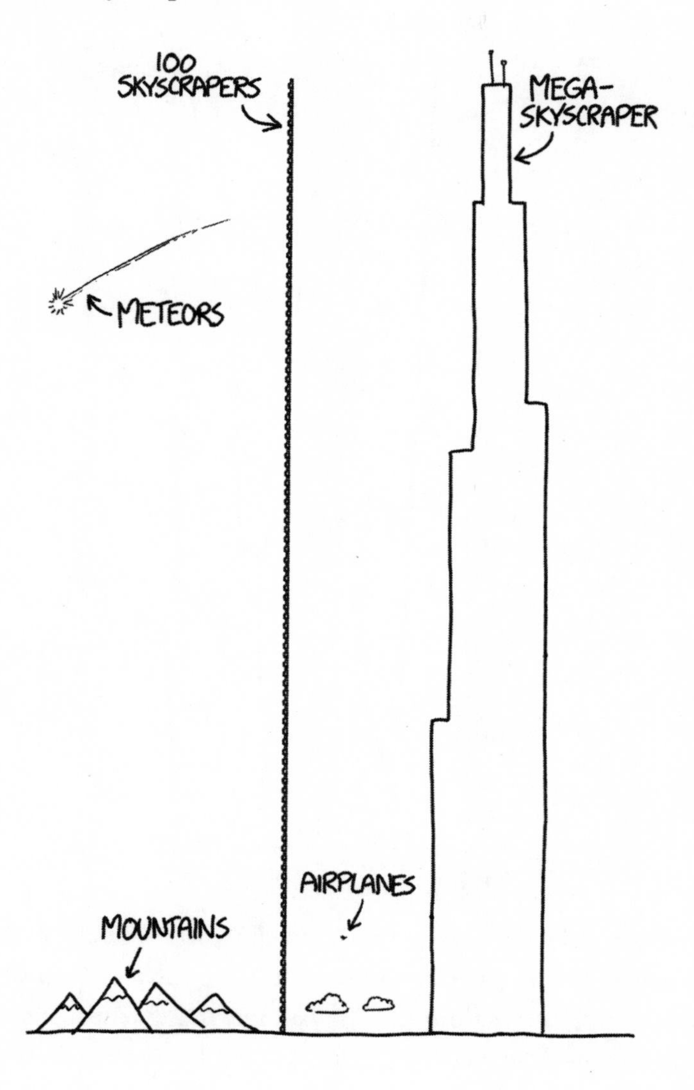

This skyscraper would still only have 10,000 floors, which is way less than your billion floors! Each of those 100 skyscrapers would have 100 floors, so the whole mega-skyscraper would have 100 times 100 = 10,000 floors.

But you said you wanted a skyscraper with 1,000,000,000 floors. Let's stack 100 mega-skyscrapers to make a mega-mega-skyscraper:

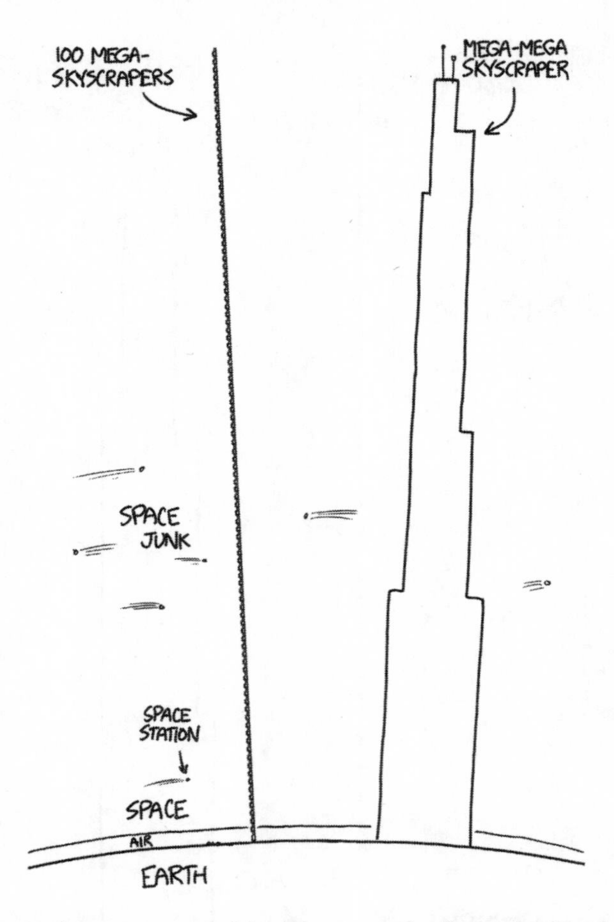

The mega-mega-skyscraper would stick out so far from the Earth that spaceships would crash into it. If the space station were heading toward the tower, they could use its rockets to steer away from it.* The bad news is that space is full of broken spaceships and satellites and pieces of junk, all flying around at random. If you build a mega-mega-skyscraper, spaceship parts will eventually smash into it.

* They'd probably get pretty grumpy after having to dodge your tower repeatedly, so you might want to launch fuel and snacks out the window with a rail gun as they go by.

Anyway, a mega-mega-skyscraper is only 100 times 10,000 = 1,000,000 floors. That's still a lot smaller than the 1,000,000,000 floors that you want!

Let's make a new skyscraper by stacking up 100 mega-mega-skyscrapers, to make a mega-mega-MEGA-skyscraper:

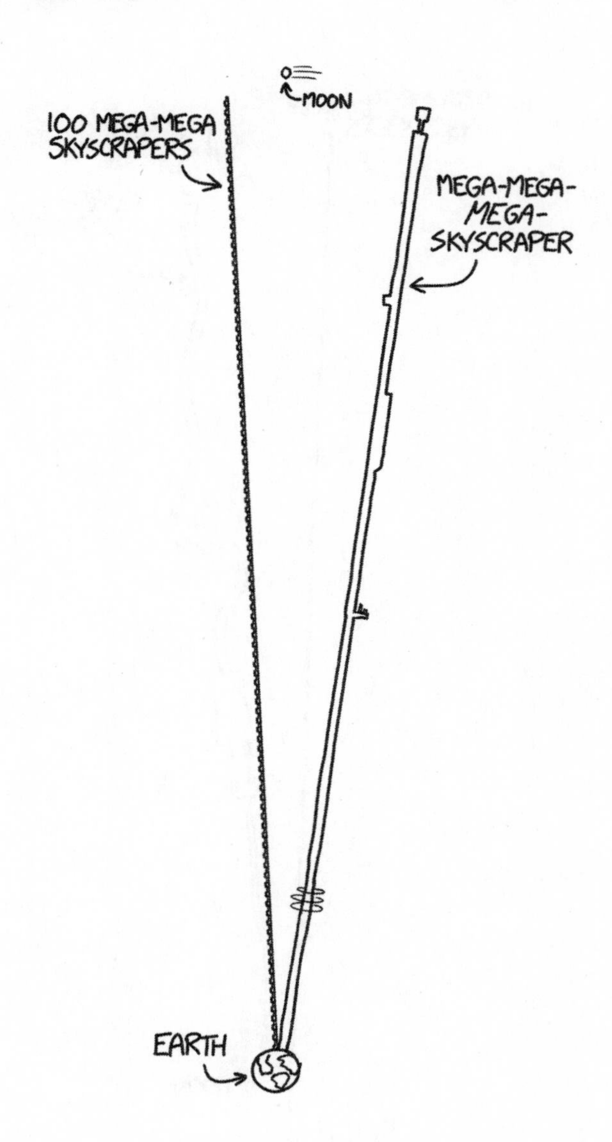

The mega-mega-MEGA-skyscraper would be so tall that the top would just barely brush against the Moon.

But it would only be 100,000,000 floors! To get to 1,000,000,000 floors, we have to stack 10 mega-mega-MEGA-skyscrapers on top of one another, to make one Keira-skyscraper:

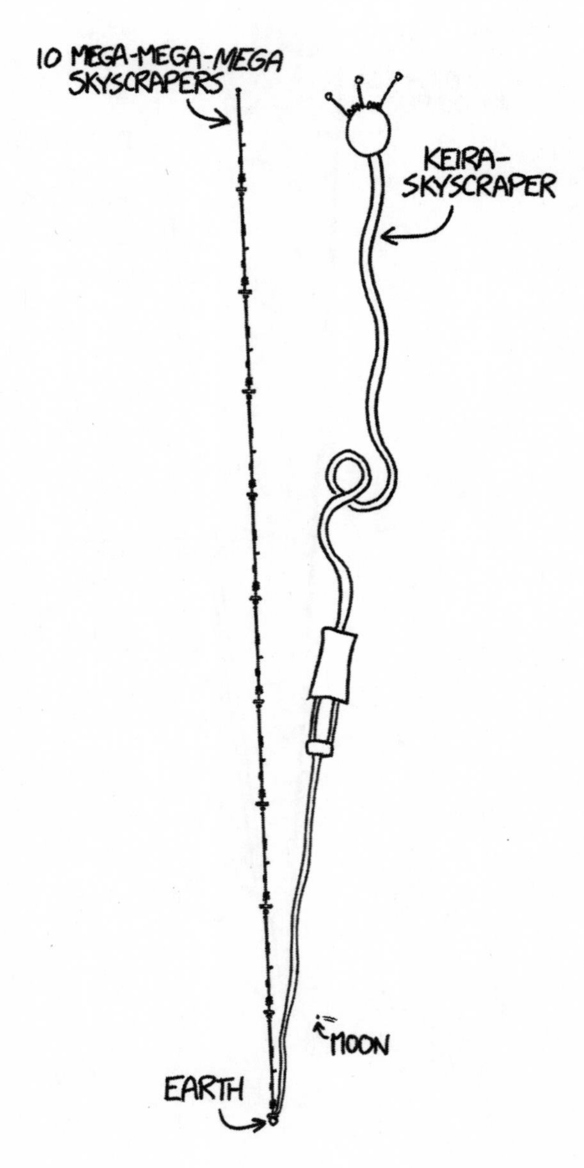

The Keira-skyscraper would be pretty close to impossible to build. You would have to keep it from crashing into the Moon, being pulled apart by the Earth's gravity, or falling over and smashing into the planet like the giant meteor that killed the dinosaurs.

But some engineers have an idea sort of like your tower—it's called a space elevator. It's not *quite* as tall as yours (the space elevator would only reach partway to the Moon), but it's close!

Some people think we can build a space elevator. Other people think it's a ridiculous idea. We can't build one yet because there are some problems we don't know how to solve, like how to make the tower strong enough and how to send power up it to run the elevators. If you really want to build a gigantic tower, you can find out more about some of the problems they're working on, and eventually become one of the people coming up with ideas to solve them. Maybe, someday, you *could* build a giant tower to space.

I'm pretty sure it won't be made of peanut butter, though.

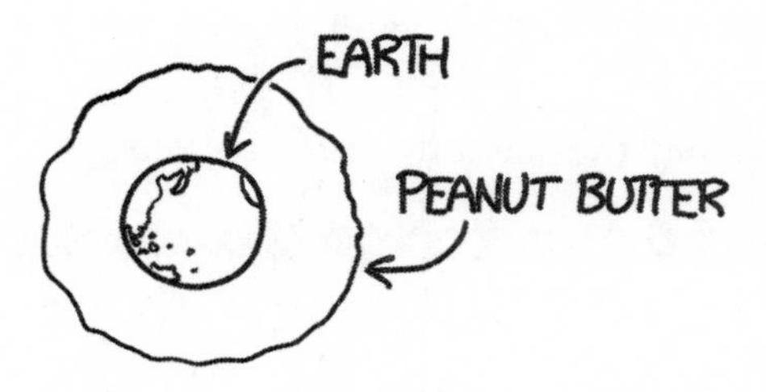

23. $2 UNDECILLION LAWSUIT

What if Au Bon Pain lost their 2014 lawsuit and had to pay the plaintiff $2 undecillion?

—Kevin Underhill

In 2014, the bakery-cafe chain Au Bon Pain (along with a few other organizations) was sued by someone demanding $2 undecillion in damages. The lawsuit was quickly dismissed, but probably not before a lot of legal folks had to look up the word "undecillion."

This is how much money the plaintiff was demanding:

$2,000,000,000,000,000,000,000,000,000,000,000,000

According to a 2021 Boston Consulting Group report, this is how much money there is in the world:

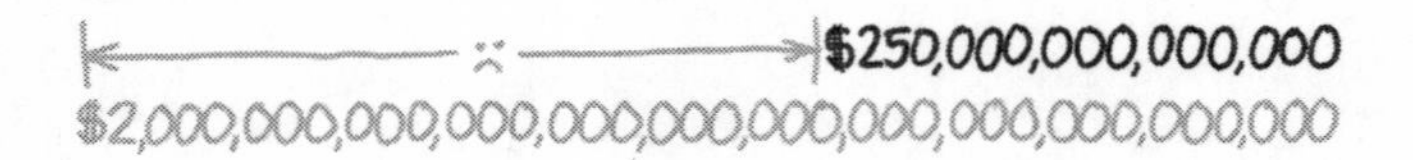

This is a rough estimate of the economic value of all goods and services produced by humanity since we first evolved:

Even if Au Bon Pain conquers the planet and puts everyone to work for them from now until the stars die, they wouldn't make a dent in the bill.

Maybe people just aren't worth enough. The EPA currently uses $9.7 million as the "value of a statistical life," although they go to great lengths to point out that this is absolutely not the value they place on any actual human life.* In any case, by their measure, the total value we place on all of the world's humans is only about $75 quadrillion.†

But people are hardly all there is on the planet. Out of all the Earth's atoms, only 1 out of every 10 trillion is part of a human. Maybe that other stuff is valuable.

The Earth's crust contains a bunch of atoms,[citation needed] some of which are probably worth something. If you extracted all the elements, purified them,‡ and sold them, the market would crash.§ But if you somehow sold them at their current market price, they would be worth . . .

|← CLOSER →|$1,600,000,000,000,000,000,000,000

$2,000,000,000,000,000,000,000,000,000,000,000,000

Oddly, that value doesn't come from things like gold and platinum. They're worth a lot, but they're rare. The bulk of the value comes from potassium and calcium, and most of the rest comes from sodium and iron. If you're going to sell the Earth's crust for scrap, those are probably the ones you should focus on.

Sadly, even selling the crust for scrap doesn't get us close to the numbers we need.

We could include the core, which is iron and nickel with a dash of precious metals, but it turns out it wouldn't help. The amount demanded in the lawsuit is just too large. In fact, an Earth made of solid gold wouldn't be enough. The Sun's weight in platinum wouldn't be, either.

---

* I can't help but notice that they don't say whether they think *that* amount would be higher or lower.

† The world's combined oil reserves are only worth a few hundred trillion, which suggests that purely from an accounting standpoint, the "no blood for oil" slogan makes a lot of sense.

‡ This is just one of many reasons that this idea wouldn't make sense in practice. The reason many elements (like ^{235}U) are valuable is that it's hard to manufacture or purify them, not just because they're rare.

§ Both in the sense that the supply would cause a drop in prices and the sense that the market is located 20 miles above the mantle and you just removed the crust supporting it.

By weight, the single most valuable thing that's been bought and sold on the open market is probably the Treskilling Yellow postage stamp. There's only one known copy of it, and in 2010 it sold for more than $2,300,000. That works out to at least $30 billion per kilogram of stamps. If the Earth's weight were entirely postage stamps, it would *still* not be enough to pay off Au Bon Pain's potential debt.*

If Au Bon Pain and co. decided to be intentionally difficult and pay their debt entirely in pennies, they would form a sphere that would squeeze inside the orbit of Mercury. The bottom line is that paying this settlement would be, in almost any sense of the word, impossible.

Fortunately, Au Bon Pain has a better option.

Kevin, who asked this question, is a lawyer and author of a legal humor blog, *Lowering the Bar,* that reported on the Au Bon Pain case. He told me that the world's most highly paid lawyer—on an hourly basis—is probably former solicitor general Ted Olson, who at one point disclosed in bankruptcy filings that he charges $1,800 per hour.

Suppose there are 40 billion habitable planets in our galaxy, and every one of them hosts an Earth-size population of 8 billion Ted Olsons.

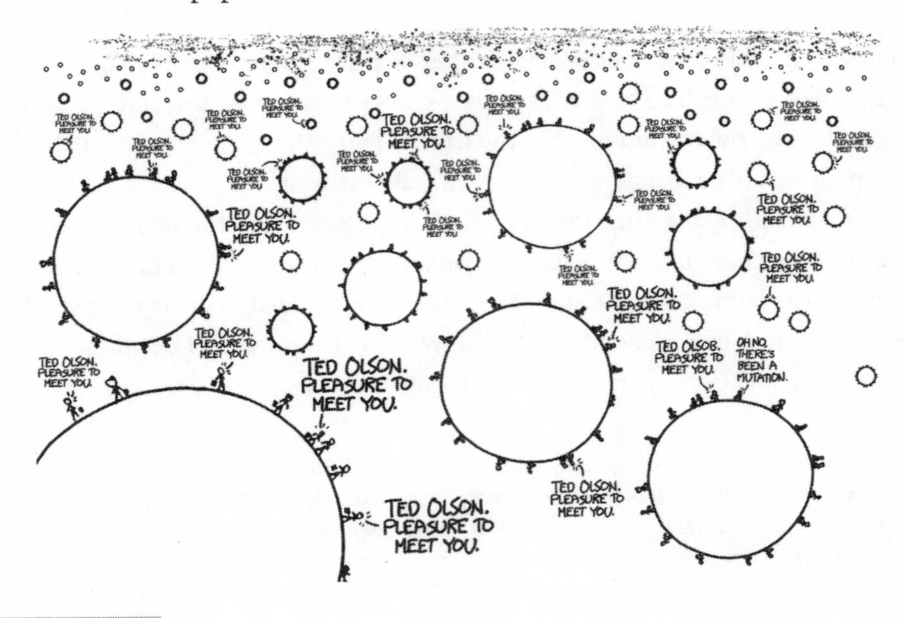

* Also, the stamps would probably be less valuable if there were literally an entire planet of them, but that's the least of Au Bon Pain's problems.

If you're ever sued for $2 undecillion, and you hired every Ted Olson in the galaxy to defend you in this case and had them all work 80-hour weeks, 52 weeks a year, for a *thousand generations* . . .

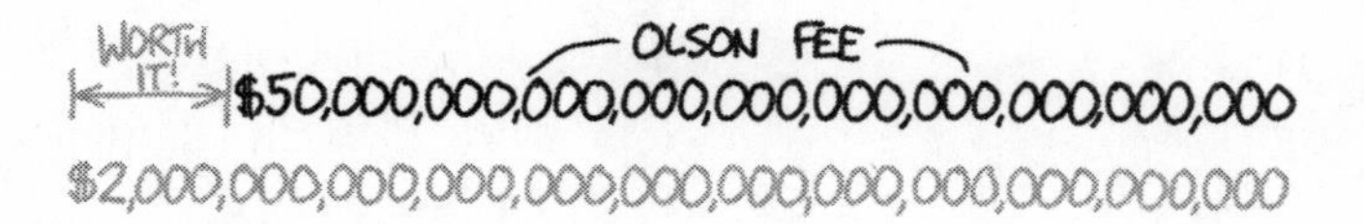

. . . it would *still* cost you less than if you lost.

24. STAR OWNERSHIP

If every country's airspace extended up forever, which country would own the largest percentage of the galaxy at any given time?

—Reuven Lazarus

Congratulations to Australia, new rulers of the galaxy.

OLD FLAG

PROPOSED NEW FLAG

The Australian flag has a number of symbols on it, including five stars that represent the stars of the Southern Cross. Based on the answer to this question, maybe their flag designers should think bigger.

Countries in the southern hemisphere have an advantage when it comes to star ownership. Earth's axis is tilted relative to the Milky Way; our North Pole points generally away from the galaxy's center.

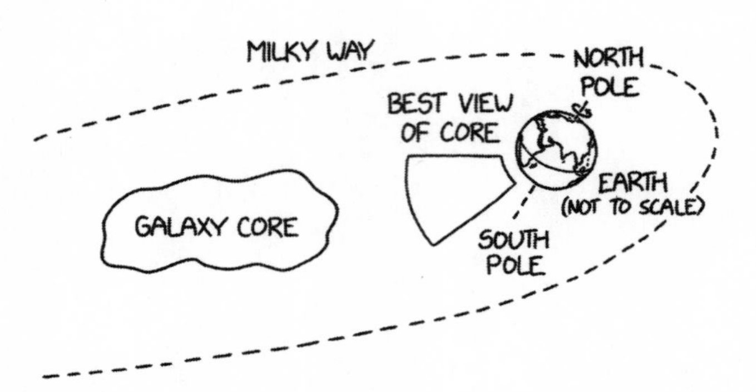

If each country's airspace extended upward forever, the core of the galaxy would stay under the control of countries in the southern hemisphere, changing hands over the course of each day as the Earth rotates.

At its peak, Australia would control more stars than any other country. The supermassive black hole at the core of the galaxy would enter Australian airspace every day south of Brisbane, near the small town of Broadwater.

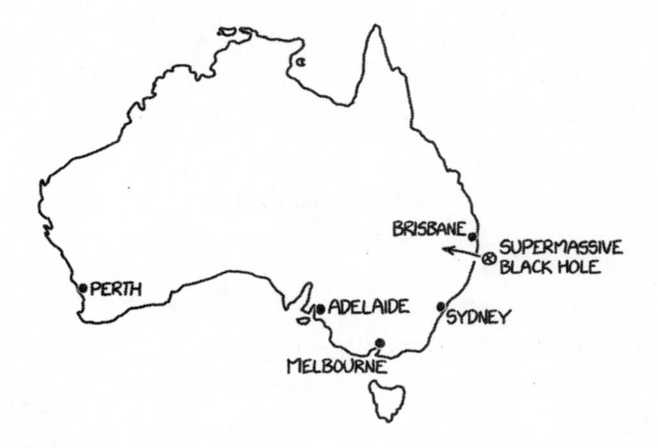

After about an hour, almost the entire galactic core—along with a substantial chunk of the disk—would be within Australian jurisdiction.

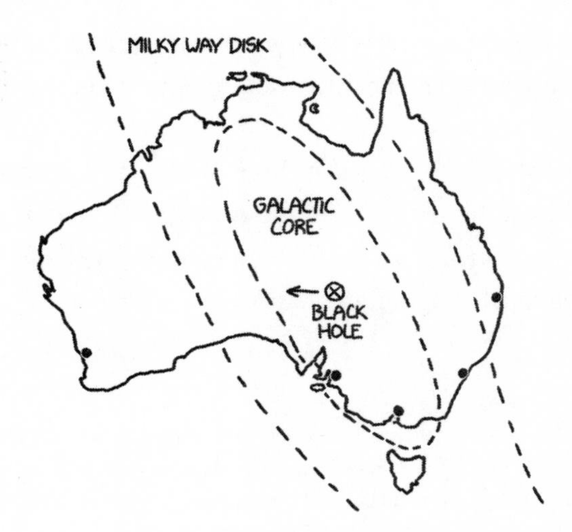

At various times throughout the day, the galactic core would pass through the domain of South Africa, Lesotho, Brazil, Argentina, and Chile. The United States, Europe, and most of Asia would have to be content with outer sections of the galactic disk.

The northern hemisphere isn't left with the dregs, though. The outer galactic disk has some cool things in it—like Cygnus X-1, a black hole currently devouring a supergiant star.* Each day, as the core of the galaxy crossed the Pacific, Cygnus X-1 would enter the United States's airspace over North Carolina.

While owning a black hole would be cool, the United States would also have millions of planetary systems constantly moving in and out of its territory—which might cause some problems.

The star 47 Ursae Majoris has at least three planets and probably more. If any of those planets have life on them, then once a day all that life passes through the United States. That means that there's a period of a few minutes each day where any murders on those planets technically happen in New Jersey.

* Cygnus X-1 was the subject of a famous bet between astrophysicists Stephen Hawking and Kip Thorne over whether it was a black hole or not. Hawking, who had spent much of his career studying black holes, bet that it wasn't. He figured that if black holes turned out not to exist, at least he would win the bet as a consolation prize. In the end, luckily for his legacy, he lost.

Luckily for the New Jersey court system, altitudes above about 12 miles are generally considered "high seas." According to the American Bar Association's Winter 2012 issue of the Admiralty and Maritime Law Committee Newsletter, this means that deaths above these altitudes—even deaths in space—are arguably covered by the 1920 Death on the High Seas Act, or DOHSA.

But if any aliens on 47 Ursae Majoris are considering bringing a lawsuit in a US court under DOHSA, they're going to be disappointed. DOHSA has a statute of limitations of 3 years, but 47 Ursae Majoris is more than 40 light-years away . . .

. . . which means it's physically impossible for them to file charges in time.

25. TIRE RUBBER

Rubber tires on millions of cars and trucks start with about ½" tread and end up bald. Rubber should be everywhere, or at least our highways should be made thicker. Where's the rubber?

—Fred

That's a good question. All that rubber has to be *somewhere*, and none of the options sound all that great.

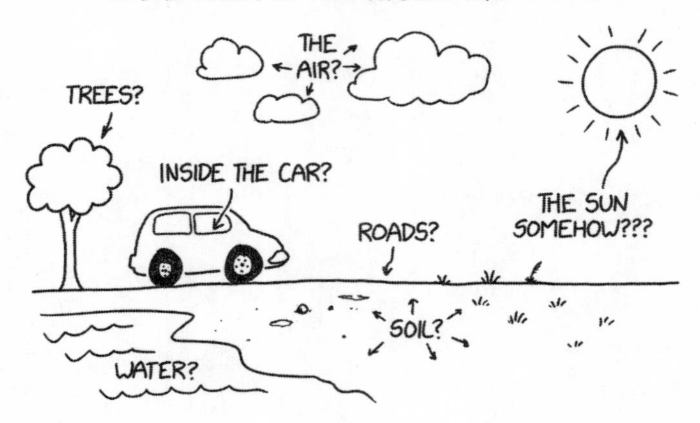

We can estimate how much rubber a tire loses—the difference between a new tire and a worn-down "bald" tire—with a simple calculation:

$$\text{Lost rubber} = \text{tire diameter} \times \text{tread width} \times \pi \times (\text{thickness}_{\text{new}} - \text{thickness}_{\text{bald}}) \approx 1.6\ \text{L}$$

That's more than a liter of rubber, which is a lot; it might be 10 to 20 percent of the total volume of the tire.

ONE LITER OF TIRE RUBBER

If a tire travels 60,000 miles before it's worn down, that means it leaves behind the equivalent of a strip of rubber about one atom thick along its path. In practice, that rubber isn't shed evenly. It comes loose in small particles and clumps, and is occasionally scraped away in large amounts all at once. If a driver slams on the brakes and skids, the tires often leave behind a stripe of rubber thick enough to see.

A lane of an especially busy highway might carry up to 2,000 cars per hour. If all the lost rubber were left behind on the lane's surface, the road would rise by about a micron per day, or a third of a millimeter per year.

It would actually be great if the tire rubber *did* stay stuck to the road, at least from an environmental point of view, but for the most part it doesn't. The particles released during ordinary driving are often small enough to drift through the air, or they get washed off the road by wind, rain, and the passage of other cars. These rubber particles waft away from the highways and end up in the air, dirt, rivers, oceans, soil, and our lungs.

I *SAID*, LUNGS ARE SUPPOSED TO BREATHE AIR!

Breathing all those tires probably isn't great for us, and it's not great for the environment, either. Tire rubber particles are a major source of microplastics in our rivers and oceans, where they affect the chemistry of the water and are often eaten by marine animals. Research into the effects of these microplastics is ongoing—for example, in 2021, a study linked salmon die-offs in the Pacific Northwest to a chemical from tire rubber in stormwater runoff.

Tire rubber waste is a tough problem to solve. We've cut back on some other sources of plastic particles in the environment—many countries have banned plastic microbeads from makeup products—but tire emissions don't seem to have a quick fix.

There are some ideas out there for reducing environmental tire rubber. Better filtering of road stormwater runoff could help. Figuring out which chemicals in tires are causing the most problems and looking for alternatives also seems like a good idea. And a few groups have proposed mechanisms for capturing rubber particles as they leave the tires.

But if you have any ideas, this is definitely an area that could use a breakthrough or two!

26. PLASTIC DINOSAURS

As plastic is made from oil and oil is made from dead dinosaurs, how much actual real dinosaur is there in a plastic dinosaur?

—Steve Lydford

I don't know.

Coal and oil are called fossil fuels because they formed over millions of years from the remains of dead organisms buried underground. The standard answer to "What kind of dead stuff does the oil in the ground come from?" is "Marine plankton and algae." In other words, there are no dinosaur fossils in those fossil fuels.

Except that's not quite right.

Most of us only see oil in its refined forms—kerosene, plastics, and the stuff that comes out of gas pumps—so it's easy to imagine the source as some uniform black bubbly material that's the same everywhere.

But fossil fuels bear fingerprints of their origins. The various characteristics of coal, oil, and natural gas depend on the organisms that went into them and what happened to their tissues over time. It depends on where they lived, how they died, where their remains ended up, and what kinds of temperature and pressure they experienced.

The dead matter carries the chemical imprint of its history—altered and jumbled in various ways—for millions of years. After we dig it up, we spend a lot of effort stripping the evidence of this story away, refining the complex hydrocarbons into uniform fuels. When we burn the fuels, their story is finally erased, and the Jurassic sunlight that was bound up in them is released to power our cars.*

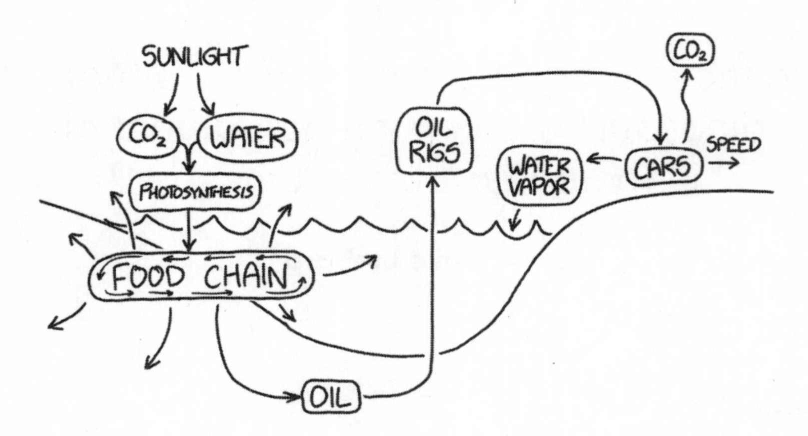

The story carried by rocks is a complicated one. Sometimes pieces are missing, discarded, or transformed in a way that misleads us. Geologists—both in academia and the oil industry—work patiently to reconstruct different aspects of these stories and understand what the evidence is telling us.

Most oil does come from ocean life buried in the seabed, which means it's mostly not dinosaurs. But the poetic idea that our fuels contain dinosaur ghosts is in some ways true as well.

There are a few things required for oil to form, including the quick burial of large amounts of hydrogen-rich organic matter in a low-oxygen environment. These conditions are most often met in shallow seas near continental shelves, where periodic nutrient-rich upwellings from the deep sea cause blooms of plankton and algae. These temporary blooms soon burn themselves out, dying and falling to the oxygen-poor

* Through photosynthesis, organisms used sunlight to bind carbon dioxide and water into complex molecules. When we burn their oil, we finally return that CO_2 and water to the atmosphere—liberating millions of years' worth of stored carbon dioxide all at once. This has some consequences.

seabed as marine snow. If they're quickly buried, they may eventually form oil or gas. Land life, on the other hand, is more likely to form peat and eventually coal.

This paints a picture like this:

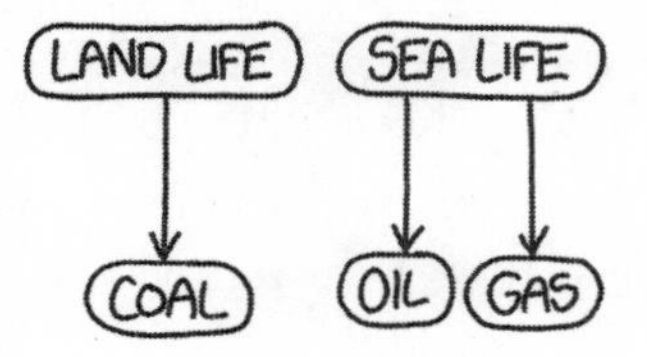

But hydrocarbon formation is a multistep process and lots of things can affect it. A huge amount of organic material washes into the ocean, and while most of it doesn't end up in oil-producing sediments, some of it does. Some oil fields—like Australia's—seem to have a lot of terrestrial sources. Most of these are plants, but some are certainly animals.*

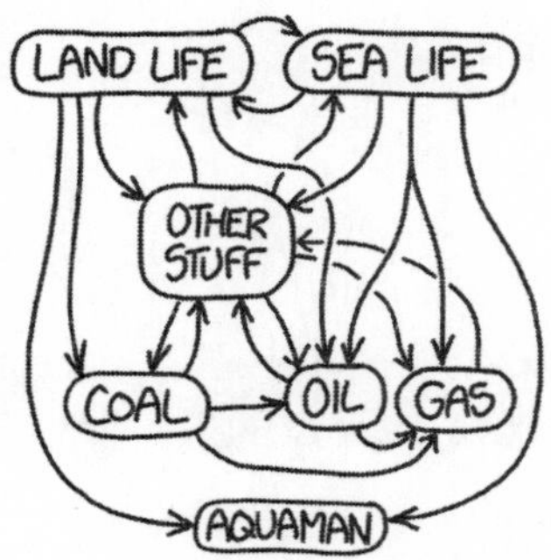

No matter where it came from, only a small fraction of the oil in your plastic dinosaur could have come directly from real dinosaur corpses. If it came from a Mesozoic-era oil field fed heavily by land matter, it might contain a slightly larger share of dinosaurs; if it came from a pre-Mesozoic field sealed beneath caprock, it might contain no dinosaur at all. There's no way to know without painstakingly tracing every step of the manufacturing process of your particular toy.

* Worth noting that while most dinosaurs were on land, a few—such as spinosaurus—were at least partly aquatic.

In a broader sense, all water in the ocean has at some point been part of a dinosaur. When this water is used in photosynthesis, molecules from it become part of the fats and carbohydrates in the food chain—but a lot *more* of that water is in your body right now in the form of water.

In other words, your plastic toys contain a lot less dinosaur than you do.

short answers 

Q How long do you think two people would have to kiss continuously before they had no lips left?

—Asli

Consider how your lips work. If lips could be worn away by pressing them against other lips, they'd already be gone.

EVER THINK ABOUT HOW YOUR TOP LIP AND BOTTOM LIP ARE KISSING?

Q My college friend and I have had this debate for years now: If you put a million hungry ants in a glass cube with one human, who's more likely to walk out alive?

—Eric Bowman

Everyone always assumes that if you put two animals together like this, they'll battle to the death, which is a very Pokémon-esque view of biology. I think both the human and the ants would be in danger more from the glass cube than from each other. And if they do get out, I think it's you and your friend who would be in danger.

Q What if all of humanity set all of their differences aside and work together to level out the Earth into a perfect sphere?

—Erik Andersen

I think you might find that the project would quickly create some new differences.

Q People talk a lot about a space elevator or a building that would reach into low orbit to save time and resources getting things into space. This is going to sound incredibly stupid, but why has no one proposed building a road into space? Since orbit is generally considered to be 62 miles out, would it be possible to build a 62-mile-high mountain somewhere in the United States? Colorado would be my suggestion, since it has a low population density and is about a mile above sea level already.

—Brian

A 62-mile mountain would have a volume of several million cubic kilometers, roughly the same amount of rock as a hundred-meter-thick slab the size of North America.

So the question is, build it out of *what*?

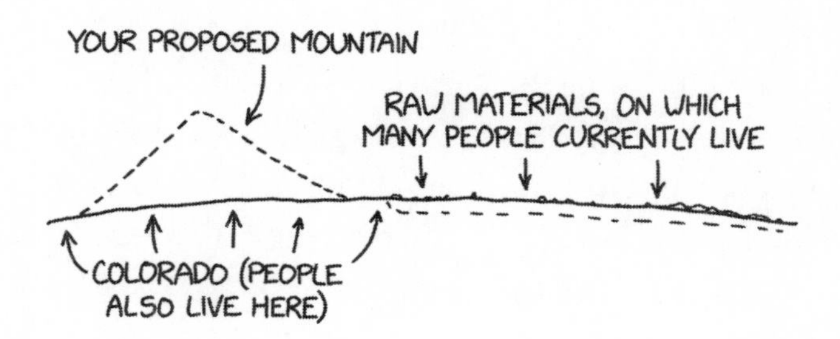

If I shot a rocket and a bullet through Jupiter's center, would they come out the other side?

—James Wilson

No.

What if Mount Everest magically turned into pure lava? What would happen to life; would we all die?

—Ian

Life would be okay.

Big piles of lava do appear on the Earth's surface every so often. These outpourings, which create massive rock slabs called "large igneous provinces," are bad news for life. There are five big mass extinctions in the fossil record, and all five of them* were accompanied by large amounts of lava blorping onto the surface.

* The dinosaur extinction, famously caused by a meteor impact in what is now Mexico, was also accompanied by one of these blorps, the Deccan traps in what is now India. The outpourings were already happening by the time the space rock arrived, though they seem to have gotten a lot worse around that time. Scientists are still debating how the two events were connected and how much each one contributed to the extinction. The main extinction seems to have happened right at the moment of impact, so it was definitely the key, but all that lava couldn't have helped the situation.

Eyes first evolved about half a billion years ago, and in that time, the Permian extinction is probably the worst thing they've seen. A large eruption of lava in what is now Siberia injected huge amounts of CO_2 into the atmosphere, causing temperatures to spike. The oceans deoxygenated and acidified. Clouds of poison gas rolled across the land. Most plant life was wiped from the continents, leaving Earth a sandy desolate wasteland. Almost everything died.

The Permian extinction involved the eruption of about a million cubic kilometers of lava. By comparison, the volume of Mount Everest—depending on how you define it—is in the thousands of cubic kilometers. Since that's pretty small compared to these large igneous provinces, your scenario probably wouldn't cause a mass extinction on the scale of the Permian.

Still, humans haven't been around that long. Even something that's 1 / 100th as bad as the Permian extinction would probably still be the worst thing that's ever happened to us. Personally, I wouldn't risk it.

Can you fall down into the Mariana Trench, or would you just swim over it?

—Rodolfo Estrella

You can do either of these things.

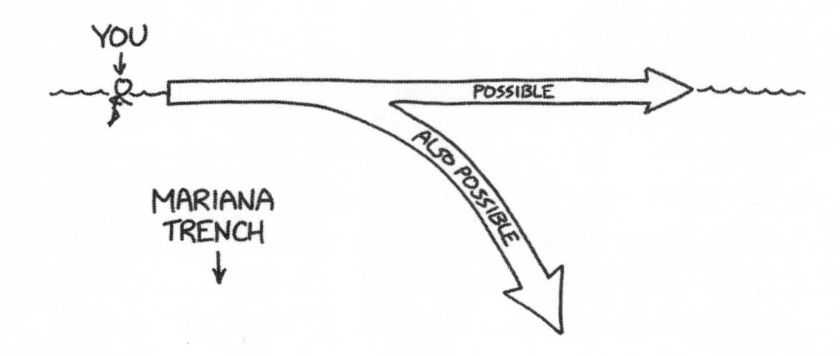

Q

I play Dungeons & Dragons, and my DM doesn't want to let us use the Gust of Wind spell to push wind into the sails of a ship and make it move. Her argument is that you can't use this spell to move a ship because someone on a sailboat can't aim a fan at the sail to propel the boat. We argue that since the spell doesn't push you backward when you use it, then we should be able to use it to make the ship sail. She says she'll allow it only if you say so.

—Georgia Paterson and Allison Adams

Of course, magic is magic, so it works however the DM says it does. That said, I take your side. If the spell doesn't push you backward when you use it, then either it's pushing off of something else or it doesn't obey the laws of physics at all. So there's no reason to expect it not to move the boat.

Besides, if the spell *does* push you backward when you use it, you can still push the boat with it. After all, fans *can* propel boats.

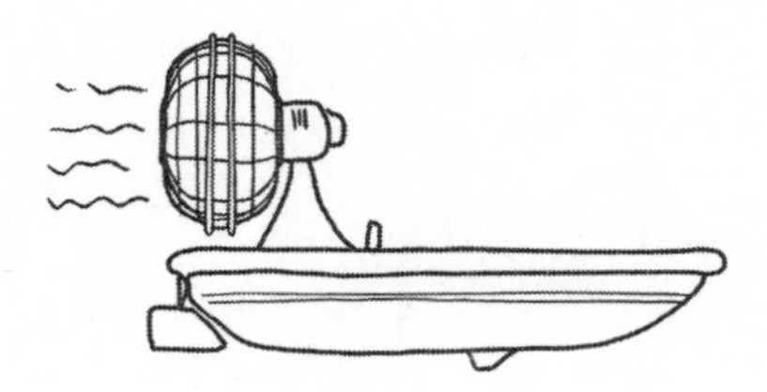

You just need to aim the spell backward.

What if I struck a match on Titan? Would it light if there's no oxygen?

—Ethan Fitzgibbon

It would spark and then be snuffed out.

Fire happens when an oxidizer—usually oxygen—reacts with a fuel. To get the reaction going, matches contain a small supply of fuel and oxidizer,* which are mixed together when the match is struck and get the reaction going. Once it does, the oxygen in the atmosphere takes over.

* The most common oxidizer used in matches, potassium chlorate, produces oxygen when heated, and is sometimes used as an emergency source of breathable air. The oxygen masks on commercial airliners are often connected to lumps of potassium chlorate. When the mask drops, a pin is pulled out, and a chemical reaction heats the potassium chlorate to produce oxygen.

On Titan, where the atmosphere is methane and nitrogen, the match would go out as soon as the oxidizer was used up.

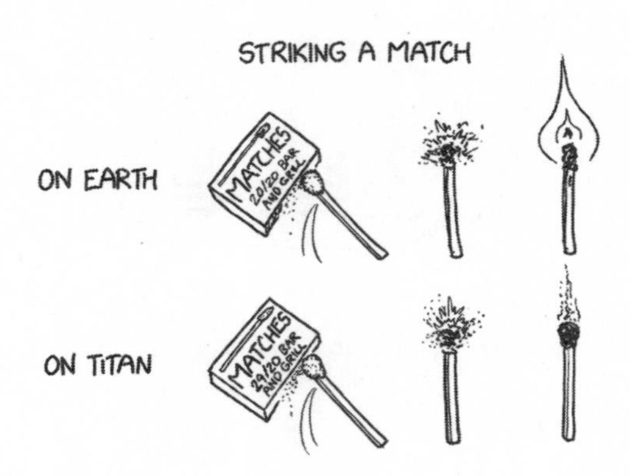

> **Q** I posted a question on social media asking what would be the smallest change that would create the biggest disaster. One of the responses I got said "if every atom gained 1 proton." So my question for you is, what would happen if every atom gained 1 proton?
>
> **—Olivia Caputo**

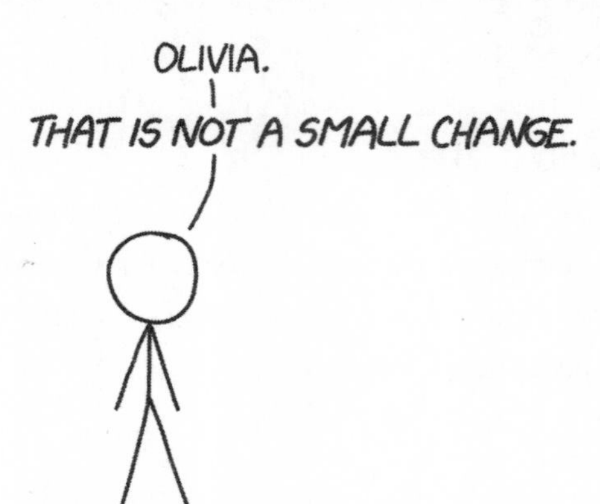

27. SUCTION AQUARIUM

When I was a child, I discovered that if I took a container into the swimming pool, I could fill it with water and then bring the container (open-end down) to the surface of the water, and the water level in my container was higher than the water level in the pool. What would happen if you tried to do this with a giant container and the ocean? Could you create a giant aquarium on top of the water that the animals could swim in and out of freely? Maybe an irregularly shaped container that you would walk around on to get closer to the fish?

—Caroline Collett

This could work.

When you lift an open-bottom container out of the water, it sucks water up with it.

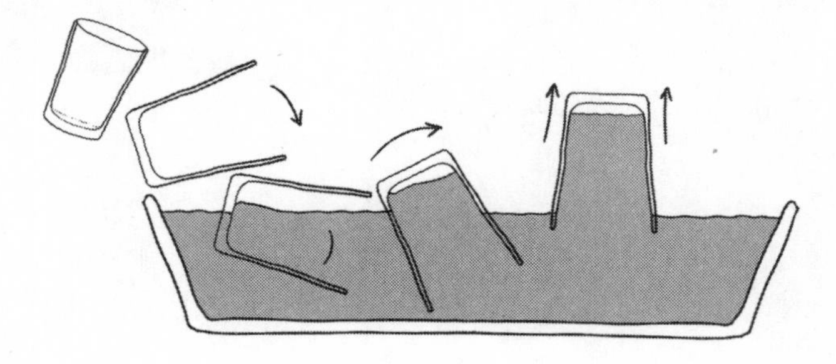

Fancy aquarium builders sometimes add raised columns like this, which they call "reverse" or "inverse" aquariums. You could put a large container in the ocean and do the same thing, giving you a raised column of seawater to look at.

Let's say you try.

You build a giant glass enclosure out of aquarium glass, put it on piles in the ocean, seal up the top, and then raise it up, lifting a meter-high column of water above the surface.

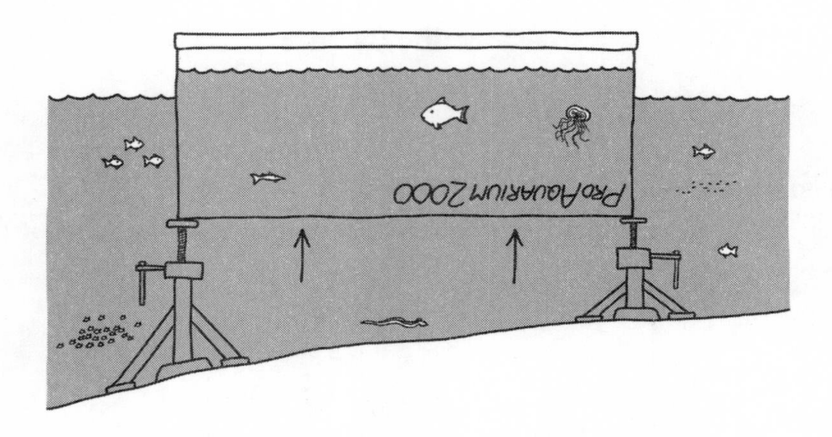

The water is held above the surface by suction—the lack of air pressure over it pushing it down. Physicists will point out that technically it's the pressure of the air on the *rest* of the ocean pushing the water up, not suction inside the column pulling it, which is true. But, just between us, once you understand that, it's still sometimes easier to think of it as suction. I think that's fine. Just don't let the physicists hear you.

Normal water is at atmospheric pressure at the surface, and higher underwater. The suction* means that the water in the column is under *less* than normal atmospheric pressure. At the surface inside the aquarium, a meter above sea level, the pressure is a little under 90 percent of one atmosphere. That's similar to the air pressure in high-altitude cities like Denver. If you swam inside and surfaced, you probably wouldn't notice the pressure difference, since your ears would be adjusting to the pressure changes from the dive anyway.

While you may not notice, the fish certainly will. Marine organisms tend to be very attentive to pressure changes since the pressure changes so quickly as you move a short distance up and down in the water. Many fish control their buoyancy through air bladders, which also help keep them upright in the water. When they ascend or descend, their buoyancy changes, and they have to change how they swim to compensate until the amount of gas in their swim bladders adjusts.

Even marine organisms without swim bladders, like sharks, notice changes in pressure. When a tropical cyclone was approaching the coast of Florida in 2001, marine biologists observed blacktip sharks heading out into the open ocean ahead of the storm, probably to escape the rough currents and pounding waves in the shallow coastal waters. Research by marine scientist Michelle Heupel and colleagues suggests that the sharks weren't responding to the wind or the waves—instead, they started their evacuation the moment they sensed the barometric pressure dropping below the normal level for the season.

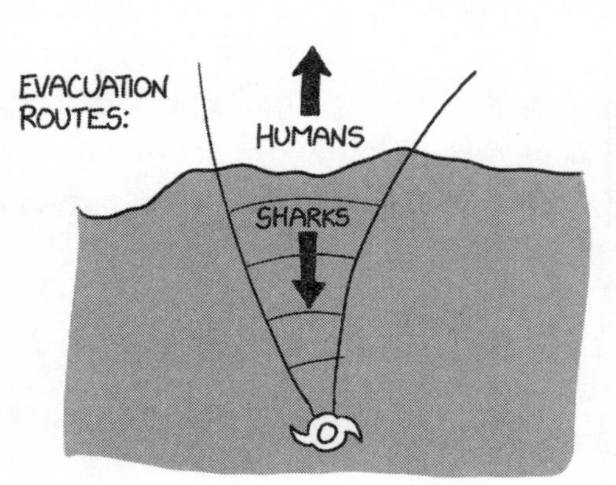

* Shhh.

Fish can survive fine at 90 percent of normal sea level pressure, so they won't have trouble swimming around in your tank, although they may be confused by the changing pressure. It wouldn't harm them, but they might mistake the pressure drop for an approaching hurricane.

A 1-meter tank is enough to see some interesting sea life, but if you want to fit really cool marine life—like the infamous white shark—you'll need to raise it higher. Your tank is barely tall enough to fit a full-size white shark's dorsal fin.

The largest exhibit at the Monterey Bay Aquarium is called Open Sea and features a 35-foot-deep tank. You might think it would be cool to raise your aquarium depth to 35 feet, giving you enough room to show off even the largest sharks.

That wouldn't work out very well.

The suction that lifts water is created by the weight of air pressing down on the ocean's surface, and air pressure isn't strong enough to lift a column of water more than about 10 meters high. By the time your column of water reached 10 meters or so, the surface wouldn't lift any higher no matter how much you lifted the enclosure. Instead, a vacuum would open up at the top and the water at the surface would start to boil in the low pressure.

If you didn't know what the air pressure in your area was, you could calculate it by looking at the height of the water in the tube. This is how many barometers work, although they usually use mercury instead of water, since mercury is much heavier and so the columns are shorter. (The mercury also doesn't boil away at the top.) When you see pressure quoted in "inches of mercury" or "mmHg," they're measuring the height of the column in a mercury suction aquarium.

Your aquarium would be a bad barometer, since the boiling water at the top would create a vapor that filled the vacuum, pushing the water down a little and giving you an inaccurate reading. But it *also* wouldn't work very well as an aquarium.

Fish that swim up the column would find that their swim bladders expanded too much, potentially causing them to rise uncontrollably. River engineers occasionally use siphons to allow water to flow over a barrier using suction, and sometimes fish swim through the tubes. When the siphons lift the fish more than 5 or 10 feet above the normal surface level, the pressure change causes serious and sometimes fatal injury, similar to the injuries caused to deep-sea fish brought to the surface too quickly.

A suction aquarium would also be perilous for any air-breathing mammals unlucky enough to swim into it. When they tried to surface, the air in their lungs would expand, potentially causing pulmonary injury if they didn't exhale. When they reached the surface, they'd find that any air remaining in the air pocket would be too thin to breathe—similar to the air on Mount Everest above the "death zone."

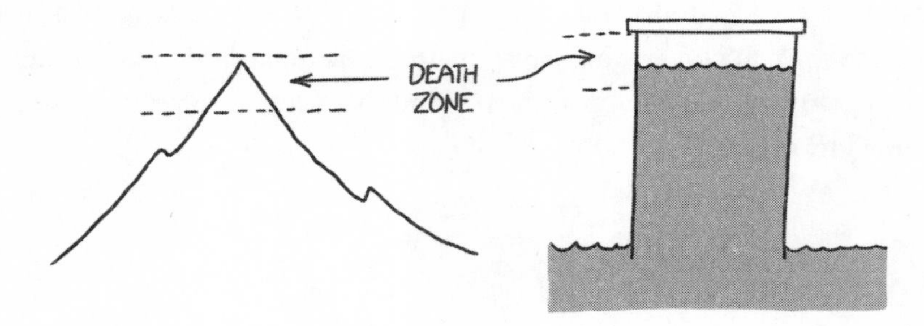

This aquarium would, thankfully, be pretty difficult to build. But it would also be temporary! If you try to build one of these tanks, you'll find that the water level drops over time. Water contains dissolved oxygen, and when the pressure is reduced, the oxygen leaves the water. In your column, dissolved oxygen would exit the water and gradually fill the space at the top of the aquarium, causing the pressure to rise and the suction effect to weaken. Over time, the water would retract back into the ocean.

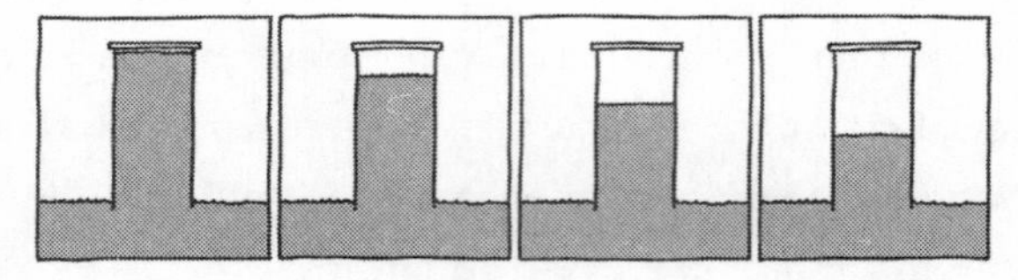

Other sources of gas could cause the water in your aquarium to drain out more quickly. Air-breathing marine mammals sometimes expel gas while swimming, and now and then a whale might swim under your aquarium.

In other words . . .

. . . your aquarium could be destroyed by whale farts.

28. EARTH EYE

If the Earth were a massive eye, how far would it see?

—Alasdair

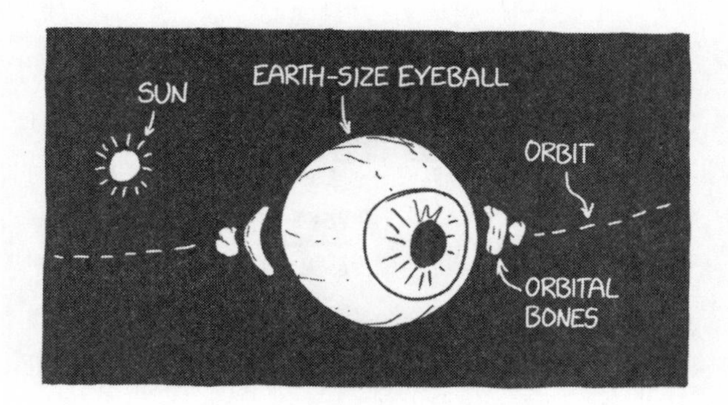

An Earth-size eyeball would have a pupil several thousand kilometers wide. A contact lens would protrude to the top of where the atmosphere should be, and a single teardrop would contain about as much water as the Earth's oceans.

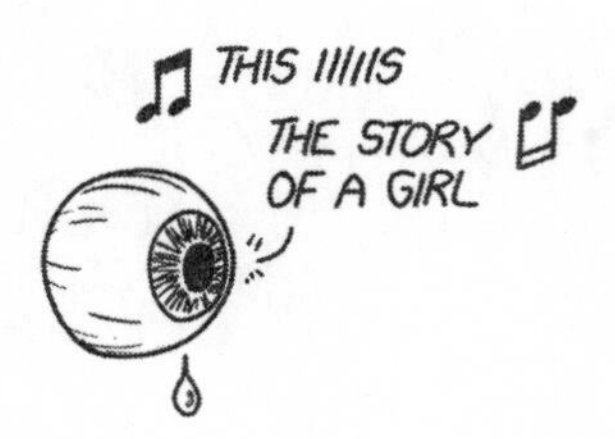

A real Earth-size eyeball wouldn't work. Light wouldn't be able to pass through that much vitreous humor, so the retina would see only darkness, and the lens wouldn't be able to hold its shape against gravity, so the eye wouldn't be able to focus. You'd also run into problems scaling up the retina—if you made the individual cells bigger, they'd no longer be able to detect visible light wavelengths.

To avoid those problems, let's imagine an Earth-size eyeball that worked like a larger version of a normal eye—with a proportionally bigger pupil and retinal area, but the same transparency and shape as a smaller one. This eyeball would be able to see incredibly well. The resolution of a telescope depends on how big the light-gathering opening is—which is why a camera with a big telephoto lens can zoom in better than your phone's camera—and the eye's huge pupil and lens would give it enormous light-gathering ability.

As long as a lens is free of defects and color distortion, the amount of detail it can see is limited mainly by diffraction, a blurring caused by the wave nature of light. This diffraction limit is proportional to the diameter of the opening.

$$\text{ANGULAR RESOLUTION} = 1.22 \times \frac{\text{LIGHT WAVELENGTH}}{\text{LENS DIAMETER}}$$

$$\text{VISIBILITY DISTANCE} = \frac{\text{TARGET FEATURE SIZE}}{\text{ANGULAR RESOLUTION}}$$

If we look at a polka-dot shirt with dots 5 centimeters apart, then we can use the visibility distance formula to calculate that if you see the shirt from more than 200 meters away, the individual dots won't be visible and it will look like the fabric is a solid color.

An eyeball the size of Earth would have a theoretical resolution half a billion times better than a normal eyeball. If it were limited only by diffraction, that eyeball would be capable of seeing whether a shirt was patterned or a solid color while it was being worn by an astronaut *on Mars*.

This telescope would theoretically be able to read a printed page of text lying on the surface of the Moon and see the shape of continents on the surface of an exoplanet orbiting Alpha Centauri.

The question "how *far* could the eyeball see" is actually pretty easy to answer—like the Webb Space Telescope, it could see just about all the way across the universe. The light from the most distant parts of the observable universe has been stretched out by the expansion of space, so most of it is shifted into the infrared, but the eyeball would be able to clearly see some of the most distant galaxies.

The eye might not be able to pick out details of those galaxies, though, thanks to the haze of space itself.

Large telescopes on Earth are limited by the turbulence of the atmosphere. Images of faraway objects shimmer and blur because air bends and distorts their light. This turbulence requires fancy adaptive optics to try to counteract, and reduces the resolution of Earth-based telescopes to below the theoretical limits of diffraction. In space, images are much sharper, so orbiting telescopes are able to operate right at those diffraction limits.

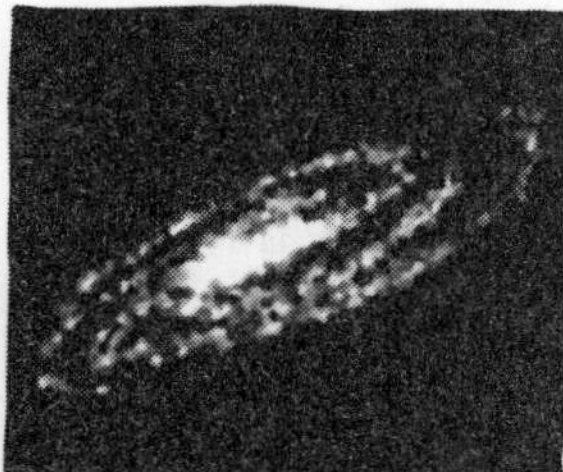

ATMOSPHERIC HAZE

To an Earth-size eyeball, space *itself* might be hazy and turbulent. A 2015 paper by astronomer Eric Steinbring suggests that quantum fluctuations in the fabric of space might distort light from distant galaxies in the same way that air distorts light from distant mountains. This distortion is too small to affect images with our current space telescopes, but it may affect larger ones, blurring the vision of an Earth-size eyeball.

Even if the things it saw were blurry, an Earth-size eyeball could see a lot farther than a regular human eye. The farthest thing a normal-size human eye can reliably see is less than 3 million light-years away—the Andromeda galaxy, or the Triangulum galaxy if you have good vision and dark skies. That's less than 0.01 percent of the distance to the edge of the observable universe. Most of the universe is too dim and far away for us to see.

The drawing below shows the Milky Way, Andromeda, and Triangulum galaxies as three dots. If you set this book on the floor in the middle of a gymnasium, the edge of the observable universe will be about as far away as the walls of the gym. When you look up at the night sky, everything you can see is inside the little circle in the center, a tiny pocket in a sprawling universe.

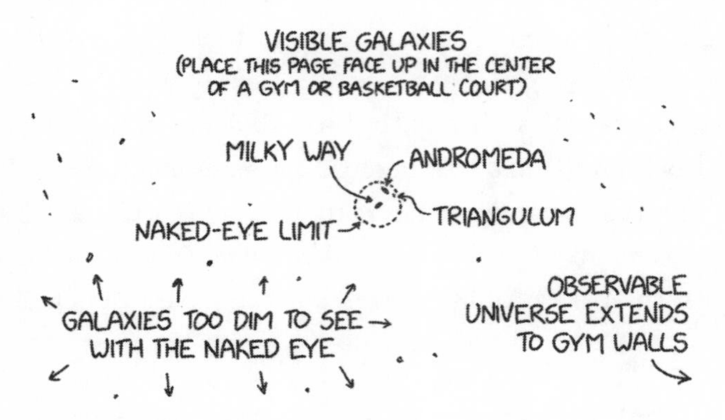

Although most of the time your vision is limited to objects within that circle, you can occasionally see much, much farther.

The night of March 18 to 19, 2008, was cloudy across much of North America, but the skies were clear in Mexico and the southwestern United States. If you had looked high in the sky at just the right time that night, you might have seen a faint dot appear for about 30 seconds in the constellation Boötes. This light was the flash from the collapse of a supermassive star about 10 billion light-years away,* thousands of times more distant than Andromeda. It set a new record for the most distant known object visible to the naked eye.

These collapsing stars emit jets of energy from their north and south poles for reasons we don't completely understand. The spin axis of GRB 080319b happened to line up directly with Earth, so we were caught right in the jet—which is why it was visible even billions of light-years away. The explosion shot a pencil-thin beam of light across the universe, like a cosmic laser pointer aimed directly at our eye.

To a human eye, the light from GRB 080319b would have looked quite faint, but to a pupil thousands of kilometers across, it could conceivably be blindingly bright. In fact, all visible stars might be too bright to look at; the focused starlight could burn the surface of the giant retina. Most people with eyes learn that it's dangerous to look directly at the Sun. But for a planet-size eye, capable of focusing so much light down to a tiny point, it might be dangerous to look at other suns, too.

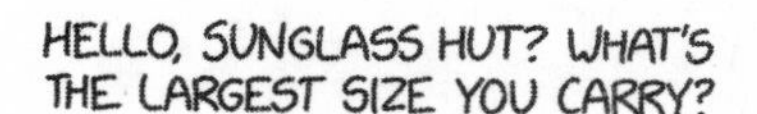

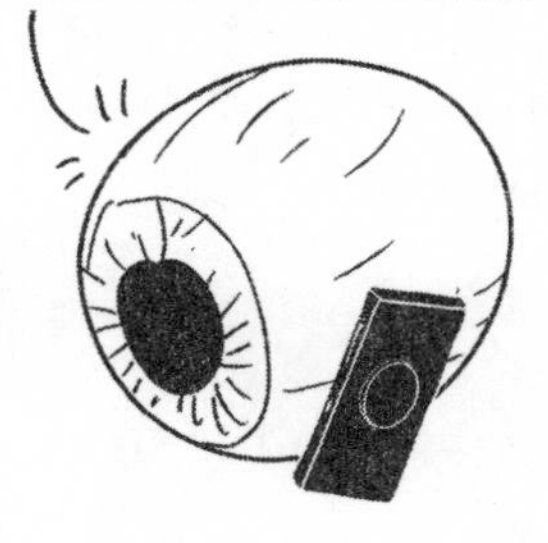

* The explosion happened about 7.5 billion years ago, but the expansion of the universe has carried it away since then, so it's more than 7.5 billion light-years away.

How many people would it take to build Rome in a day?

—Lauren

The number of people isn't necessarily the bottleneck. Like the old joke goes, it takes a person nine months to produce a baby, but assigning nine people to the job won't make it take one month. If you send more and more people to build Rome, at some point you'll just have a chaotic and disorganized mess.

A series of studies in the 1990s and 2000s by civil engineer Daniel W. M. Chan and colleagues used data on construction in Hong Kong to come up with formulas for how long construction projects will take to finish based on their overall cost and physical size.

For the purposes of a very rough estimate, looking at GDP and property values for cities of similar sizes suggests that the total value of all property in Rome might be about $150 billion. If we assume—again, this is a very rough estimate—that construction costs are about 60 percent of market value, that puts the cost to construct Rome at about $90 billion.* If we plug that into Chan's formula, it suggests that Rome should take 10 to 15 years to construct. We'd need to speed that up by a factor of 5,000 or so if we wanted to finish it in a day.

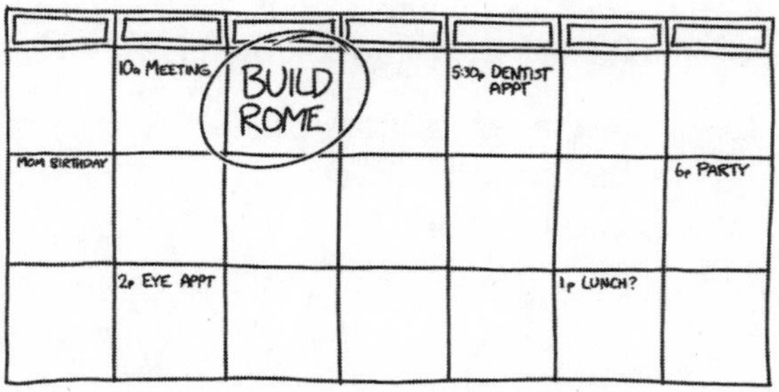

Adding more people can only speed that up so much. At some point, the main bottleneck will be training and coordinating everyone to avoid massive traffic jams as supply trucks bring in people and material. They say all roads lead to Rome, which would be helpful if it were true, but a glance at a map shows that a lot of roads are on totally different continents.

But let's suppose that we could assemble the entire world's population† and that we could solve all the training, coordination, and traffic problems—considering only

* Looking at a few American cities, we can see that the total value of all property in a region tends to be a little larger than the annual GDP of the region. For example, the combined value of all property in Cook County, Illinois, (Chicago) was assessed at about $600 billion in 2018, when the county's GDP was $400 billion. New York City has about $1.6 trillion of property and a GDP of $1 trillion. Rome's GDP is a little more than $100 billion, which suggests the total value of all property might be something like $150 billion.

† Gathering the whole world's population together in one place would be a bad idea, as discussed in the *What If?* chapter "Everybody Jump." Rome has an area of 1,285 square kilometers, which means we'd be packed in at mosh-pit-like densities of six or seven people per square meter. That would be too dense a crowd to comfortably stand around in, let alone to do construction work.

labor. How quickly could we build Rome? Let's try a few different ways of estimating the answer and see how well they agree.

My friends recently installed a new tile floor in their bathroom and the cost of labor for the tile installation was about $10 per square foot. Let's assume—and I know this sounds like a stretch, but bear with me here—that a city is the same as a tile floor. Rome has an area of 1,285 square kilometers, which means it would cost $140 billion to tile the whole place, at least if they got the same contractor my friends did.* If the world charges $20/hour for labor, then that's 7 billion hours. With 8 billion people on the job, that means we should be able to knock it out in just under an hour.

Let's try a different approach. If we use our GDP-based estimate of $90 billion for Rome's construction cost, and if 30 percent of construction cost is labor, then at $20/hour it should take a little over 2 billion hours of labor to construct Rome. With 8 billion people, that comes out to 15 minutes—a little faster than our tile estimate, but still in the same general range.

TIME TO BUILD ROME

MODEL	RESULT	VS ACTUAL HISTORY
BATHROOM TILE METHOD	50 MINUTES	25,000,000x FASTER
ROUGH GUESS FROM GDP	15 MINUTES	90,000,000x FASTER

* If Rome's municipal government wants a quote, I can put them in touch.

Of course, it's silly to model a city full of monuments, historic works of art, and priceless treasures as if it's a tile floor or a modern apartment building. So let's come at it from the other direction.

The ceiling of the Sistine Chapel is among the world's most famous and iconic works of art. Michelangelo spent 4 years creating the sprawling series of paintings, covering an area of 523 square meters.*

If we assume that Michelangelo painted for 40 hours a week, 52 weeks a year, then he painted at a rate of 1 square meter every 16 hours. At that rate, it would take 20 billion Michelangelo-hours to cover all of Rome with a city-size Renaissance masterpiece. Divided among 8 billion people, that's just 2½ hours of labor, or 150 minutes.

TIME TO BUILD ROME

MODEL	RESULT	VS ACTUAL HISTORY
BATHROOM TILE METHOD	50 MINUTES	25,000,000x FASTER
ROUGH GUESS FROM GDP	15 MINUTES	90,000,000x FASTER
SISTINE CHAPEL METHOD	150 MINUTES	9,000,000x FASTER

That's not wildly different from the half-hour estimate we come up with by modeling the city as a tile floor, and once again it suggests that building Rome in a day is not as implausible as it seems, from a labor standpoint.

Of course, you can't build Rome in a day. First of all, it's already been built, so the people there would get mad if you tried to build it again. And even if you built it somewhere else, you wouldn't be able to fit everyone in the required space, get them the materials they needed to build their portion, and keep everyone on task and on schedule.

* Painters like to say that he could've finished it in a weekend if he'd used a roller.

You'd face organizational problems beyond simply deciding who does what tasks. The Sistine Chapel is in Vatican City, which is within—but not technically part of—Rome, so it's not clear whether it will be included in Lauren's construction project. If it is, then the work of painting the chapel's ceiling will be divided up among thousands of different people.

Expect some artistic clashes.

30. MARIANA TRENCH TUBE

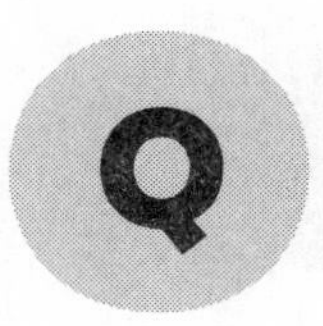

If I put an indestructible 20-meter-wide glass tube in the ocean that goes all the way down to the deepest part of the ocean, what would it be like to stand at the bottom? Assuming the Sun goes directly overhead.

—Zoki Čulo, Canada

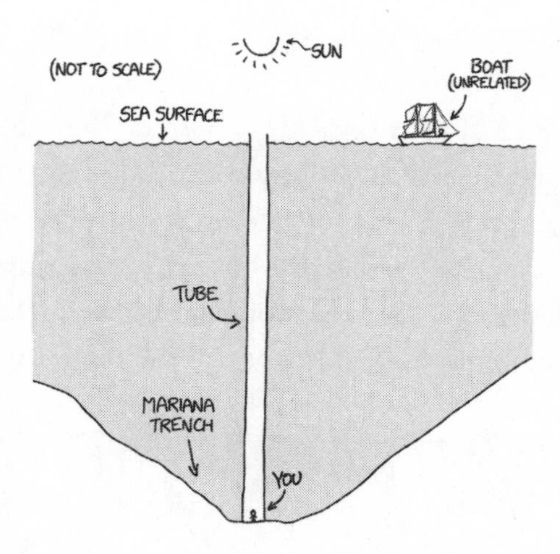

Your tube would be three times deeper than the deepest mines. In deep mines, it's hot and the air pressure is high. In your tube, heat wouldn't be a problem; the heat in mines comes from the rocks, which get hotter as you go deeper. The deep ocean is barely above freezing, so the walls of your tube would be cold, keeping the air cool.

The air pressure in the tube would be very high, several times that of the surface. This pressure wouldn't have anything to do with the high-pressure water around you, which would be held back by the tube. The pressure would be high because you're so far below sea level. Air pressure doubles with every six kilometers you descend, so at a depth of 10 kilometers, it would be nearly four times higher than what you're used to. Luckily, humans can handle that kind of pressure change without too much trouble—hyperbaric chambers, used for treating certain medical conditions, subject people to similar pressures. Just make sure to ascend slowly to avoid decompression sickness.

The Sun will only pass directly over the mouth of the tube on a few days each year, around April 20 and August 23. On those days, for a minute or two, you would actually be able to see just fine! Even though only a small portion of the sun would be visible, the sun is very bright,[citation needed] so the bottom of the tube would be as bright as a well-lit room. The dense air above you would absorb and scatter a little more light than usual, slightly dimming the Sun, but not really enough to notice.

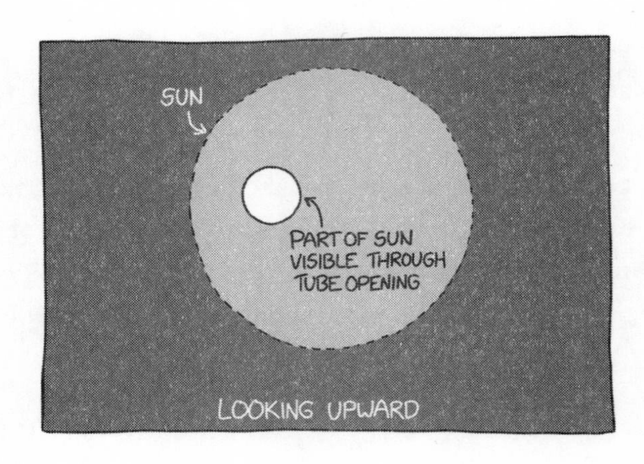

The water around you would be dark. If you shone a flashlight through the wall, you'd most likely see empty expanses of silt, but you might spot the occasional critter such as a sea cucumber. If you do, you should take notes; only a few people have visited the bottom of the trench, so we don't know what sort of life is most common there.

After the Sun passes overhead, you'll be stuck in pitch-black darkness for another 6 months, so you'll probably want to hop in your elevator and return to the surface.

If you don't have an elevator, you could always try returning to the surface the fun way: make a hole in the side of the tube and wait.

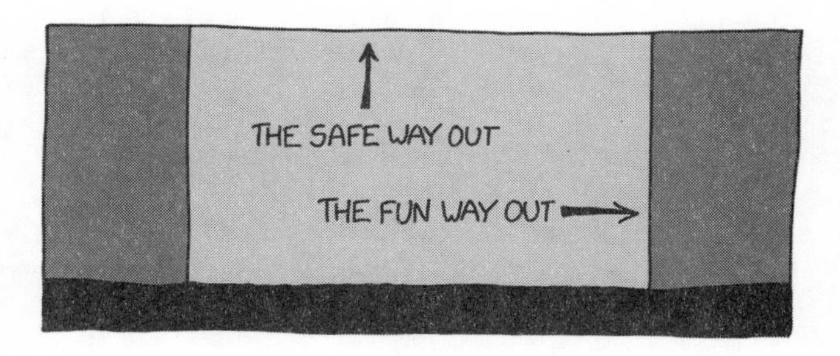

If you decide to cut a hole in the side of the tube, don't stand in front of it. The enormous water pressure of the Challenger Deep would propel a supersonic jet of water through the opening.

If you opened the bottom of the tube completely and let the sea flood in freely, the column of water would rush up it at Mach 1.3. If you tried to ride this jet upward, you wouldn't survive the violent acceleration from the initial impact of the water. To ascend safely, you'd need to let the tube start to fill up in a slower and more controlled fashion.

Once the first kilometer or two of the tube was filled, you could fully open the bottom of the tube without experiencing dangerously violent accelerations. If you had some kind of a giant plunger to keep all the water below you, the acceleration would rocket you up out of the tube in under a minute. When you reached the opening at the top, you would be traveling at 500 miles per hour, and would be carried high above the ocean's surface on a fountain of icy water.

Surprisingly, the fountain of water might keep running once you were done. In 1956, oceanographer Henry Stommel suggested that, because of differences in temperature and salinity between the surface and the deep ocean, if you connect the surface and the deep ocean with a tube and push water through it, it might continue flowing indefinitely.

The tube wouldn't create perpetual motion. The steady flow is possible because the surface and depths of the ocean aren't quite in equilibrium, thanks to a subtle imbalance in how temperature and salinity equalize between them. Since the water inside the tube can equalize temperature with its surroundings through the wall of the tube but not exchange salinity, Stommel's calculations showed the tube could upset the equilibrium and cause the ocean to mix. A 2003 experiment with a PVC tube over the Mariana Trench (not extending all the way to the bottom!) confirmed that this effect could cause a slow exchange of water.

Some people have suggested this could be used to cool the ocean surface to weaken hurricanes, fertilize the water with deep-sea nutrients to encourage growth, or dispose of waste. Stommel himself was skeptical. He ended his 1956 paper by commenting, "It seems premature to speculate upon the improbable practical importance of this phenomenon," and noted that "as a power source it is quite unpromising. Thus it remains essentially a curiosity."

31. EXPENSIVE SHOEBOX

What would be the most expensive way to fill a size-11 shoebox (e.g., with 64 GB Micro SD cards all full of legally purchased music)?

—Rick

The limit for the value of a shoebox seems to be about $2 billion. Surprisingly, this turns out to be true for a wide range of possible fillings.

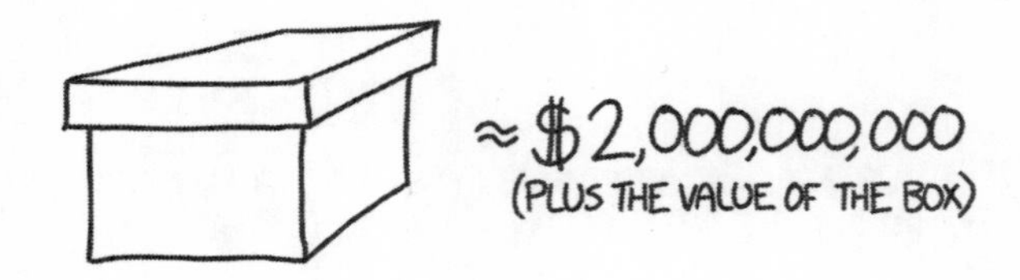

Micro SD cards are a good idea. Say you fill them with songs that cost about $1 each, and Micro SD cards have a capacity of about 1.6 petabytes per gallon. A men's size-11 shoebox is about 10 to 15 liters, depending on the brand and type of shoe, which means it can hold up to 1.5 billion four-megabyte songs. (Or 1.5 billion copies of one song, if you have an artist you *really* want to support.)

Expensive enterprise software can have a slightly higher cost-to-megabyte ratio, since it often retails for thousands of dollars and takes up gigabytes of space.

Once you start considering software prices, you can probably crank the "cost" of things in a shoebox as high as you want by getting involved in cryptocurrency or making unlimited in-app purchases in some pay-to-play mobile game. And while the resulting RPG character on your phone may technically represent the result of your spending that much money, it's hard to argue with a straight face that your character is in any sense *worth* a trillion dollars.

So let's think about actual objects.

There's gold, of course. Thirteen liters of gold is worth about $14 million as of 2021. Platinum is a little more expensive at $16 million/shoebox, about 10 times the value density of $100 bills. On the other hand, a shoebox full of gold would weigh as much as a small horse, so it might not be as practical as $100 bills if you're trying to go shopping.

There are more expensive metals. A gram of pure plutonium, for example, would cost about $5,000.* As a bonus, plutonium is even denser than gold, which means you could fit almost 300 kilograms of it in a shoebox.

Before you spend $2 billion on plutonium, take note: Plutonium's critical mass is about 10 kilograms. You could technically fit 300 kilograms of it in a shoebox, but you could only do so *briefly*.

---

* At least, as best as I can tell from some internet searches. In other news, I'm now on a lot of government watch lists.

High-quality diamonds are expensive, but it's hard to get a handle on their exact price because ~~the entire industry is a scam~~ the gemstone market is complicated. InfoDiamond.com quotes a price of more than $200,000 for a flawless 600 milligram (3 carat) diamond—which means that a shoebox full of perfect-quality gem diamonds could theoretically be worth $15 billion—although since you'd have to pack several smaller diamonds together to fill the shoebox tightly, $1 or $2 billion might be more reasonable.

Many illegal drugs are, by weight, more valuable than gold. Cocaine's price varies a lot, but in many areas it is in the neighborhood of $100/gram.* Gold is currently less than half that. However, cocaine is much less *dense* than gold,† so a shoebox full of cocaine would be less valuable than one full of gold.

* Update: I'm now on the rest of the government watch lists.

† But wait—what *is* the density of cocaine? I spent a while reading through a wonderfully earnest and citation-filled discussion of this question on the Straight Dope Message Boards where several people tried to get to the bottom of this question. They were able to determine cocaine's boiling point and solubility in olive oil, but in the end gave up on figuring out the density and just decided it was probably about 1 kilogram/liter, like most organic substances.

Cocaine is not the most expensive drug in the world. LSD, which is sold by the microgram, costs about a thousand times more than cocaine by weight—it's one of the only substances commonly purchased in microgram increments. A shoebox full of pure LSD would be worth about $2.5 billion. The active ingredients in vaccines are also often measured in micrograms, so even though they're not that expensive per dose, a shoebox worth of mRNA or influenza virus protein would also be worth billions.

At the other end of the price-per-dose spectrum, some prescription drugs aren't particularly small, but they are extremely expensive. A dose of brentuximab vedotin (Adcetris) can cost $13,500, putting its shoebox value in the same $2 billion range as LSD, plutonium, and Micro SD cards.

Of course, you could always put *shoes* in the shoebox.

Judy Garland's shoes from *The Wizard of Oz* sold at auction for $666,000, and—unlike the other things we've considered—may have, at one point, actually been placed in a shoebox.

If you really want to fill a shoebox with an arbitrarily large amount of money, you could get the US treasury to mint you a trillion-dollar platinum coin, something that—due to a loophole in a law around minting commemorative coins—it is technically authorized to do.*

But if you're open to leveraging our monetary system's legal authority to impart value into an arbitrary inanimate object . . .

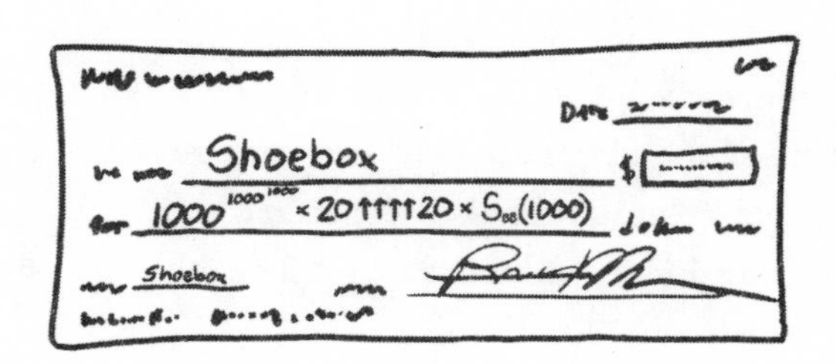

. . . you could just write a check.

* Hopefully, by the time you're reading this, this loophole is still just weird trivia.

32. MRI COMPASS

Why don't compasses point toward the nearest hospital because of the magnetic fields created by MRI machines?

—D. Hughes

They do, and it can be a problem!

MRI medical scanners have powerful magnets in them. The scanners are shielded, so the strongest parts of the magnetic field are contained inside the scanner, but a weaker field extends out around it. This "fringe field" falls off quickly as you move farther from the machine, but its influence can still be felt some distance away.

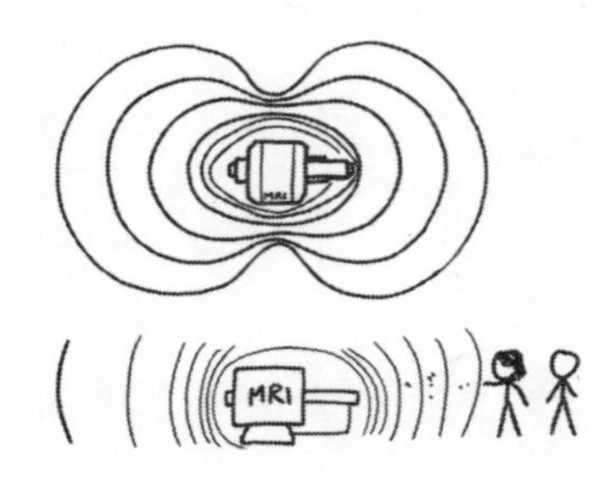

The manual for one popular MRI scanner says that, to prevent the fringe fields from causing damage, certain sensitive objects should be kept away from the machine. It suggests that credit cards and small motors should be kept 3 meters away, computers and disk drives should be kept 4 meters away, pacemakers and X-ray tubes should be kept 5 meters away, and electron microscopes should be kept 8 meters away.

If you try to walk toward Earth's magnetic north pole using a compass, the fringe field from an MRI could deflect you from your path, but only if you get close enough. The Earth's magnetic field strength varies from place to place, but it's generally somewhere between 20 and 70 μT. The fringe field from an MRI scanner falls below this level at a distance of about 10 meters, so that's roughly the maximum distance at which you could capture someone navigating by compass.

The captured person's path would curve away from the north pole of the MRI magnet and toward the south pole:

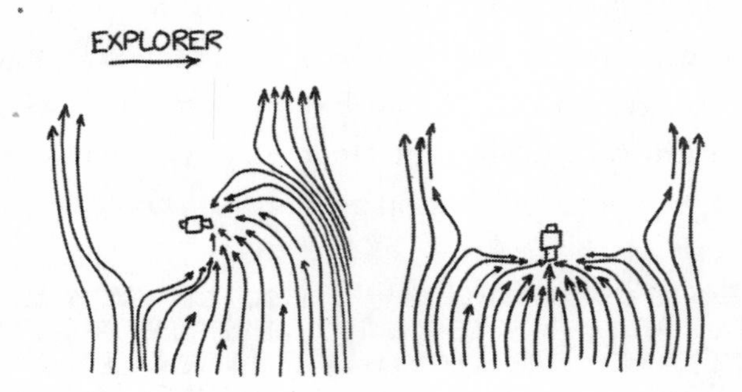

It might seem confusing that someone navigating toward Earth's north pole would be attracted to the MRI's south pole, but that's because the Earth's pole names are backward. The "north" end of a magnet is the one that points toward the Earth's north pole, which means the Earth's north magnetic pole is technically a south magnetic pole, and vice versa. This is deeply annoying to me, but there's nothing we can do about it, so we might as well move on.

If someone in the middle of North America were trying to walk toward the north magnetic pole, and you tried to capture them by placing a random MRI scanner somewhere in Canada, the chances that they would be deflected by it would be around 1 in 500,000. According to the Canadian Medical Imaging Inventory, there were 378 MRI scanners in operation in Canada in 2020, which means that by scattering them across Canada you could create a magnetic net* that would capture roughly 1 in 1,300 pole-seeking explorers. The other 1,299 would reach the actual magnetic north pole, so even with hundreds of MRIs, this would be a pretty ineffective explorer capture method.

EXPLORERS WHO WOULD BE DIVERTED BY AN MRI SCANNER

EXPLORERS WHO WOULD REACH THE MAGNETIC NORTH POLE (OR DIE OF EXPOSURE)

* Or "magnet" for short.

But this whole scenario isn't quite as unrealistic as it sounds.

While the magnetic fields from MRI machines aren't strong enough to lure in compass-guided explorers from around the country, they *have* occasionally played similar tricks on a smaller scale.

A 1993 report by the US Department of Transportation described an incident in which a medical helicopter was coming in to land on a hospital's rooftop landing pad. As the helicopter approached the ground, the magnetic compass and some related equipment suddenly indicated that the helicopter had unexpectedly rotated 60 degrees. Luckily, the pilot was able to ignore the faulty instrument readings and land safely. The culprit turned out to be an MRI scanner in a trailer that was parked near the helipad.

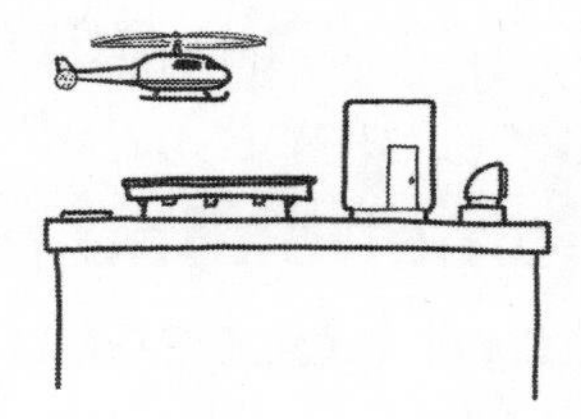

So you don't need to worry that some distant MRI scanner will influence your compass navigation through the forest. But if you're landing a helicopter near a hospital, definitely keep an eye out.

33. ANCESTOR FRACTION

I noticed recently that the number of people in a family tree increases exponentially with each generation: I have 2 parents, 4 grandparents, 8 great-grandparents, and so on. Which got me thinking—are most people descended from the majority of Homo sapiens who have ever lived? If not, what fraction of all the people that have ever lived am I descended from?

—Seamus

You're not descended from most humans who have ever lived. You're probably descended from about 10 percent of them, although it's going to be hard to get an exact number.

The average person has two parents and—excluding periods of global population decline—an average of at least two children. That means that our ancestors and descendants *both* tend to grow exponentially. As you count backward or forward through time, the set of people you're related to grows. Every child links up two family trees, and every lineage that survives for more than a few generations tends to grow exponentially until it includes everyone.

Our set of ancestors grows the same way. Each of your ancestors represents the merger of two family trees, so more and more people are included as you move further back. It's possible for your family tree to occasionally shrink as you trace it backward in time—for example, if you had a group of ancestors who were isolated for many generations—but it never dies out. If you follow it back far enough, the relentless doubling means that eventually you'll reach a date at which all surviving lineages have been absorbed into your family tree. At that point, all the people who left descendants are your ancestors, and you and everyone else have the same set of ancestors.

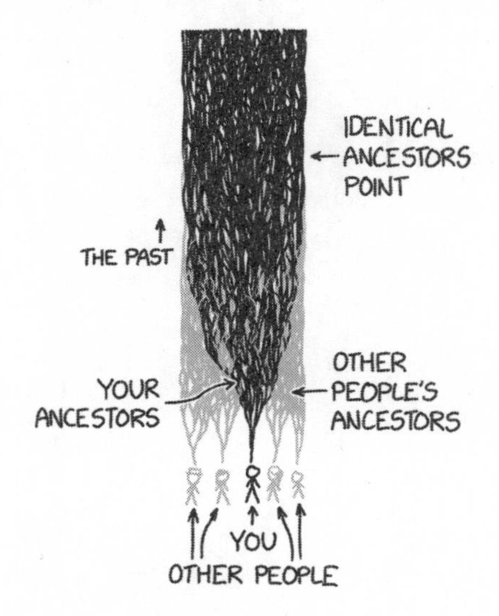

A 2004 simulation by Douglas L. T. Rohde and colleagues estimates that the identical ancestors point is likely somewhere between 5000 and 2000 BCE. At that date, everyone who left descendants at all is an ancestor of everyone. Each lineage from that date has either died out or expanded to include all living humans, and so all living humans share the same set of ancestors from that point backward.

The majority of people who have children end up contributing to this family tree. Rohde et al. estimate that in human populations, 60 percent of humans who had any children at all ended up in the family tree permanently, and 73 percent of people who survived to adulthood had children. If we assume 55 percent survive to adulthood, based on studies of historical child mortality, then that implies about 25 percent of all humans who were ever born went on to have children and leave a permanent lineage of descendants.

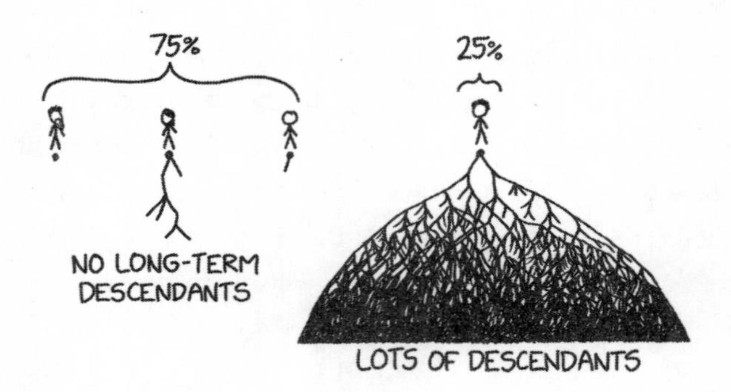

Combining that number with historical population and birth rate estimates suggests that about 20 billion humans lived before the identical ancestors point, which means about 5 billion are your ancestors.

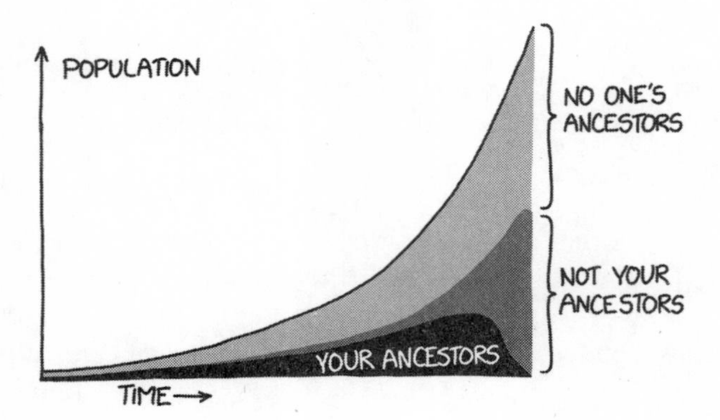

After the identical ancestors point, your set of ancestors no longer overlaps exactly with everyone else's, but it still includes a lot of people. Before the identical ancestors point, your family tree resembles a braided stream. Only in the final millennium or so does it shrink to resemble a tree. Over this time, you probably add another 5 to 10 billion ancestors.

All in all, your family tree likely includes 10 to 15 billion humans out of the 120 billion or so who have ever lived. That means that, under a modern calendar, 33 million of them have a birthday today.

Unless it's February 29.

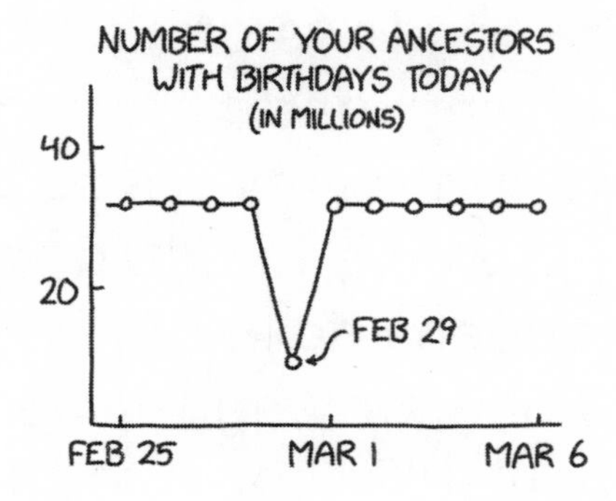

I'm a lowly college student stuck in a car without AC. As such, the windows are down most of the time when I'm driving, and I started thinking: If a bird happens to match my speed and direction perfectly, and I swerve to catch the bird in my car . . . what happens next, other than an angry bird? Does the bird stay right where it was? Fly into the windshield? Drop into the seat? My roommate and I disagree. Any help settling this would really make all our lives easier.

—Hunter W.

This is the kind of thing that seems like it wouldn't work, but as much as it pains me to say it, it honestly might. The bird would definitely be confused and angry, but if you somehow caught it by surprise and successfully executed this maneuver, it would probably come through the capture intact. Congratulations on your new pet bird.

Let's look at what happens at the moment you swerve to engulf the bird with your car.

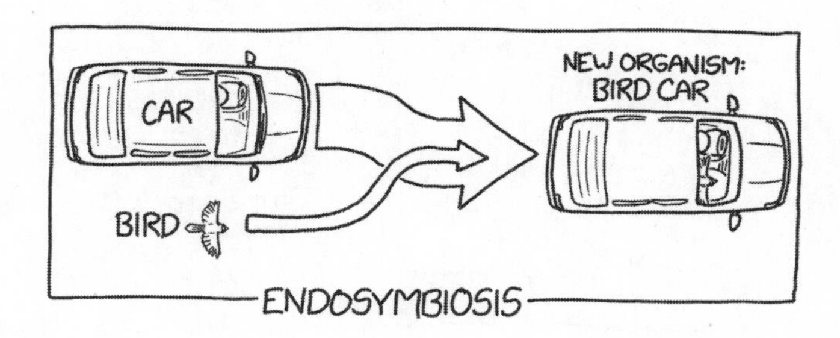

Suppose you and the bird are both traveling at 45 mph. When you swerve to engulf it, you'll both still be traveling at 45 mph—the bird will just be inside the car. From the bird's point of view, it's been flying into a 45 mph headwind as you hover alongside it.

To fly at a steady speed, birds flap their wings.[citation needed] A fast-moving bird experiences a lot of drag, which they counteract by flapping to produce thrust.

The air inside the car is moving at 45 mph. When the bird passes through the window, the headwind the bird has been flying into will abruptly vanish. Without that drag, the flapping will be producing thrust, so if it kept flapping, the bird would start accelerating forward relative to the car—like if you were running on a treadmill and the belt abruptly stopped moving.

A broad-winged hawk flying at an airspeed of 45 mph experiences about a third of a newton of drag, which means their flapping needs to produce a third of a newton of thrust to counteract it.* If the hawk kept flapping the same way once the drag was gone, that thrust would start accelerating it forward.

If all other forces stayed the same, that third of a newton of thrust would be enough to gradually accelerate the hawk toward the front of the car at 1 m/s^2, causing it to bonk gently into the windshield within a second or two. But all other forces wouldn't stay the same.

Without the headwind rushing past, the hawk's wings would no longer provide lift, and it would suddenly find itself dropping. Gravity would accelerate it downward at 9.8 m/s^2, much more than the 1 m/s^2 forward acceleration from the continued flapping.

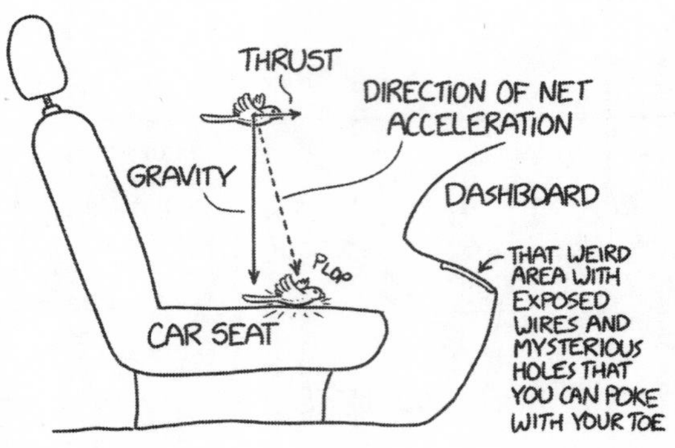

The two forces would combine to send the hawk plopping down onto the passenger-seat cushion.

But we're neglecting one big, big factor, which is how the bird would react. Most birds don't *want* to take a road trip with you.[citation needed] Birds that are startled frequently try to take off and fly toward what looks like open space, which is often how they end up hitting windows. If the window is close enough, the bird won't have time to get up enough speed to injure itself too badly, which is why the Audubon Society recommends that if you can't put bird feeders more than 10 meters away from your window, you should put them *closer* than 1 meter.

The windshield in your car might be too close for the bird to really injure itself, but it certainly wouldn't be good for the bird to fly into it. You said you keep your

* This explains why migrating hawks soar instead of flapping all the time—flapping for eight hours would consume their entire daily metabolic budget.

windows down, so hopefully in this unlikely situation the bird would manage to find its way back out without injury.

If the bird doesn't *want* to leave the car, that's a different problem entirely, and you should probably contact a wildlife rehabilitator for help.

Unless the bird is just tired of flying everywhere. Maybe it would appreciate a ride for once.

35. NO-RULES NASCAR

If you stripped away all the rules of car racing and had a contest that was simply to get a human being around a track 200 times as fast as possible, what strategy would win? Let's say the racer has to survive.

—Hunter Freyer

The best you'll be able to do is about 90 minutes.

There are lots of ways you could build your vehicle—an electric car with wheels designed to dig into the pavement on turns, a rocket-powered hovercraft, or a pod that runs along a rail on the track—but in each case, it's pretty easy to develop the design to the point where the human is the weakest part.

The problem is acceleration. On the curved parts of the track, drivers will feel powerful g-forces. The Daytona International Speedway in Florida has two main curves, and if the vehicles went around them too fast, the drivers would die from the acceleration alone.

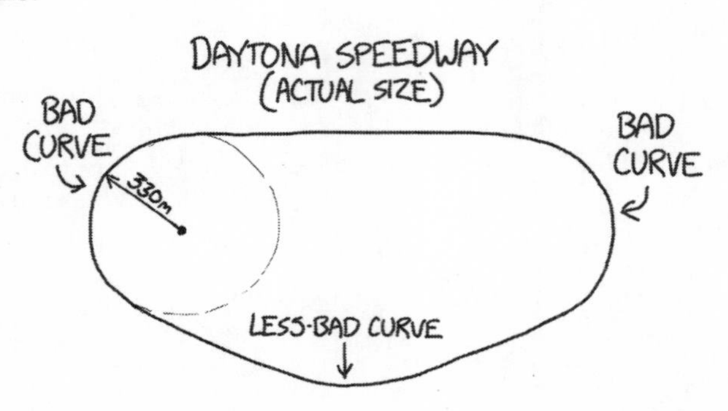

For extremely brief periods, such as during car accidents, people can experience hundreds of Gs and survive. (One G is the pull you feel when standing on the ground under Earth's gravity.) Fighter pilots can experience up to 10 Gs during maneuvers, and—perhaps because of that—10 Gs is often used as a rough limit for what people can handle. However, fighter pilots only experience 10 Gs very briefly. Our driver would be experiencing them in pulses, for minutes and probably hours.

Since rocket launches involve a lot of sustained acceleration, NASA has compiled extensive research on human acceleration tolerance. But the most fun data comes from an Air Force officer named John Paul Stapp. Stapp strapped himself into a rocket sled and pushed his body to the limit, taking careful notes after every run. He was a memorable character; an article on his experiments by Nick T. Spark in *Wings & Airpower Magazine* includes the line " . . . Stapp was promoted to the rank of major [and] reminded of the 18 G limit of human survivability . . ."

Stapp's brief experiments with extremely high accelerations notwithstanding, most data shows that for periods on the order of an hour, normal humans can only handle 3 to 6 Gs of acceleration. If we limit our vehicle to 4 Gs, its top speed on the turns at Daytona will be about 240 mph. At this speed, the course will take about 2 hours to complete—which is definitely faster than anyone has driven it in an actual car, but not *that* much faster.

But wait! What about the straightaways? The vehicle will be accelerating during the turns but coasting on the straightaways. We could instead accelerate the vehicle up to a higher speed while on straight segments, then decelerate it back down when approaching the end. This would result in a speed profile like this:

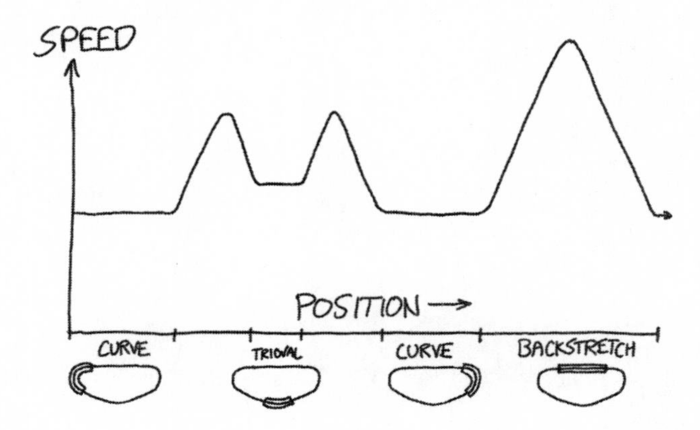

This kind of variable-speed path has the additional advantage that—with some clever back-and-forth maneuvering on the track—the driver can be kept at a relatively constant acceleration through the whole trip, hopefully making the forces easier to endure.

Keep in mind that the direction of the acceleration will keep changing. Humans can survive acceleration best if they're accelerated forward, in the direction of their chest, like a driver accelerating forward. The body is least capable of being accelerated downward toward the feet, which causes blood to pile up in the head.

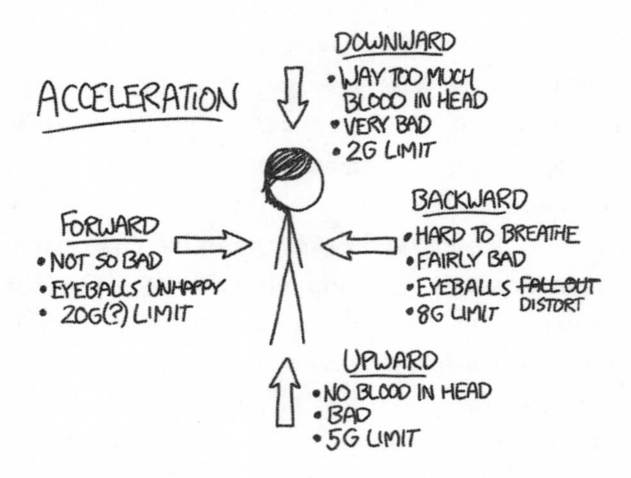

To keep our driver alive, we'll need to swivel them around so they're always being pressed against their back. (But we have to be careful not to change direction too fast, or the centrifœtal* force from the swiveling of the seat will itself become deadly!)

* I've gotten tired of the arguments about "centrifugal" versus "centripetal" and decided to split the difference.

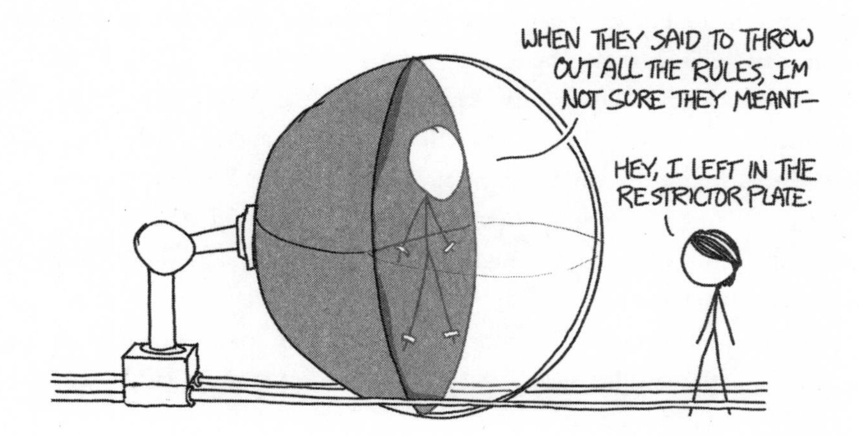

The fastest modern Daytona racers take about 3 hours to finish the 200 laps. If limited to 4 Gs, our driver will finish the course in a little under 1 hour and 45 minutes. If we raise the limit to 6 Gs, the time drops to 1 hour and 20 minutes. At 10 Gs—well past what humans could tolerate for an extended period—it would still take an hour. (It would also involve breaking the sound barrier on the backstretch.)

So, barring dubious and untested concepts like liquid breathing—filling the lungs with oxygenated fluid to allow us to withstand higher accelerations—human biology limits us to Daytona finishing times over an hour.

What if we drop the "survive" requirement? How fast can we get the vehicle to go around the track?

Imagine a "vehicle" anchored with Kevlar straps to a pivot in the center of the track, balanced with a counterweight on the other side. In effect, this would be a giant centrifuge. This lets us apply one of my favorite weird equations, which says that the edge of a spinning disk can't go faster than the square root of the specific strength* of the material it's made of. For strong materials like Kevlar, this speed is 1 to 2 km/s. At those speeds, a capsule could conceivably finish the race in about 10 minutes—although definitely not with a living driver inside.

Okay, forget the centrifuge. What if we build a solid chute, like a bobsled course, and send a ball bearing (our "vehicle") rocketing down it? Sadly, the disk equation

* (tensile strength divided by density)

strikes again—the ball bearing can't roll faster than a couple km/s or it will be spinning too fast and will tear itself apart.

Instead of making it roll, what if we make it slide? We could imagine a diamond cube sliding along a smooth diamond chute. Since it doesn't need to rotate, it could potentially survive more accelerations than a rolling ball bearing. However, the sliding would result in substantially more friction than the ball-bearing example, and our diamond might catch fire.

To defeat friction, we could levitate the capsule with magnetic fields and make it progressively smaller and lighter to accelerate and steer it more easily. Oops—we've accidentally built a particle accelerator.

And while it doesn't exactly fit the criteria in Hunter's question, a particle accelerator makes for a neat comparison. The particles in the Large Hadron Collider's beam go very close to the speed of light. At that speed, they complete 500 miles (30 laps) in 2.7 milliseconds.

There are probably about a thousand motor-racing tracks in the world. The LHC beam could run the equivalent of a full Daytona 500 on each of those tracks, one after another, in about 2 seconds, before the drivers had made it to the first turn.

And that's *really* as fast as you can go.

weird & worrying #2

Q What would happen if you put the end of a vacuum hose up to your eye and turned on the vacuum?

—Kitty Greer

Q Is it possible to hold your arm straight out of a car window and punch a mailbox clean off its pole? Could you do it without breaking your hand?

—Ty Gwennap

THIS IS GOING TO HURT YOU MORE THAN IT'S GOING TO HURT ME.

Q If people's teeth kept growing, but when they were fully grown they come off and are swallowed, how long would it take before it causes any problems?

—Valen M.

THIS QUESTION HAS ALREADY CAUSED ME A PROBLEM.

Q In a defensive situation, how much epinephrine (in an EpiPen) would it take to subdue a possible attacker?

—Henry M.

DON'T WORRY—THE EPIPEN IS MIGHTIER THAN THE SWORD.

36. VACUUM TUBE SMARTPHONE

What if my phone was based on vacuum tubes? How big would it be?

—Johnny

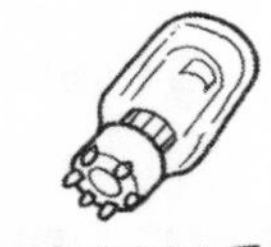

In principle, any computer built from transistors can be built from vacuum tubes, and vice versa.

Transistors and vacuum tubes use different mechanisms to do the same basic task: If they receive an electrical signal, they flip a switch one way, and if they don't, they flip it the other way. That switch controls some *other* electrical signal, which can be used to tell *other* switches what to do. We build digital circuits by chaining these parts together, creating complicated sets of rules for taking in inputs and producing outputs.

In his 1937 master's thesis, mathematician Claude Shannon showed how vacuum tubes could be arranged to implement any set of logical steps, providing a blueprint for how to build Alan Turing's universal computer using practical electrical components. Transistors replaced vacuum tubes by the 1960s, because the transistor was much smaller and more reliable, but the same digital circuits can be built from either of them.

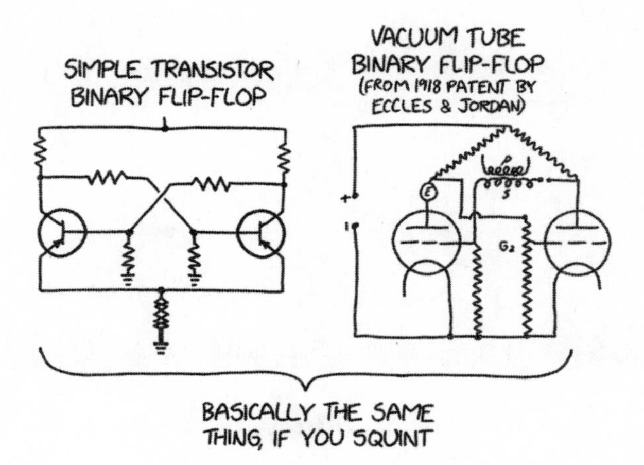

Early computers were extremely large by modern standards. ENIAC, the first programmable computer, was taller than a person and 30 meters long. UNIVAC, a commercial computer built a few years later, was a more compact cube shape, but was still the size of a room.

A modern smartphone is smaller than ENIAC or UNIVAC, but it has a *lot* more digital switches. UNIVAC had a little over 5,000 vacuum tubes packed into its 25 m^3 case. An iPhone 12 has 11.8 billion transistors packed into the phone's 80 mL case, which is about a trillion times more computer per liter.

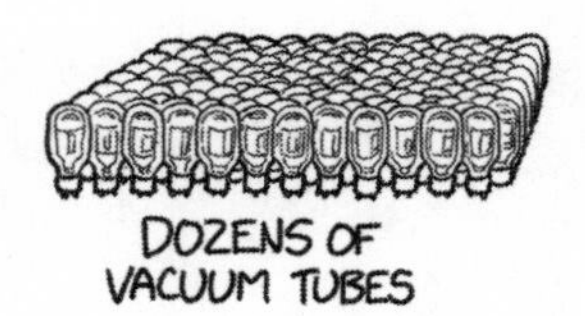

If you built an iPhone with vacuum tubes instead of transistors, packed together with the same density as they were in UNIVAC, the phone would be about the size of five city blocks when resting on one edge.

Conversely, if you built the original UNIVAC out of iPhone-size components, the entire machine would be less than 300 microns tall, small enough to embed inside a single grain of salt.

The vacuum tubes themselves wouldn't take up all that space. If you could build all the other parts of the VacPhone using modern components, you could make the whole thing smaller. A common vacuum tube from the early days of computing was the 7AK7, which was about the size of a piece of sidewalk chalk; 11.8 billion 7AK7s packed together into the shape of an iPhone would fit in a single city block.

Your phone would have some problems. One is that it wouldn't run very fast. Digital circuits perform steps one after another, with the transition from one step to the next coordinated by a clock. The faster the clock runs, the more steps the computer can perform per second. Vacuum tubes are actually fairly good at high-speed switching, but UNIVAC still only used a 2 MHz clock, about 1/1,000th the speed of modern computers.

Your phone would be so big that you'd have to worry about the speed of light. It would take signals so much time to travel from one end to the other that the different parts of the phone would be out of sync with one another. If your phone was running at 2 MHz, when the clock at one end ticked, the signal from that tick wouldn't have time to reach the other end of the phone before the next tick started.

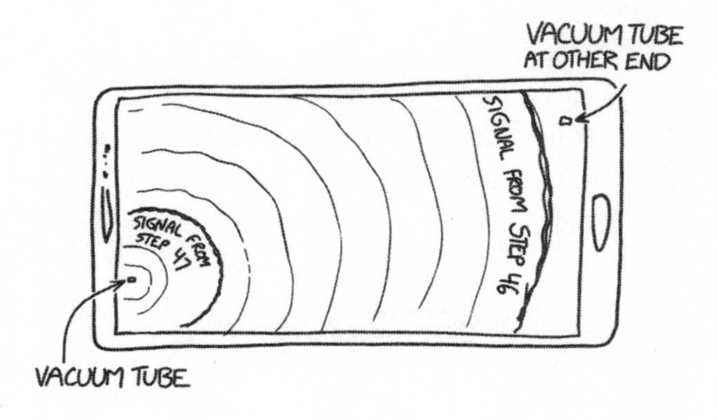

The sluggish speed of light means that you'd have to arrange the components of your phone to work in parallel as much as possible. That way, a computation at one end wouldn't be stuck waiting for a result of a computation at the other.

This sounds ridiculous, but modern computers have exactly this problem. If a chip is running at 3 GHz, light—and electric signals—don't have time to cross from one end of the computer to the other during a single clock cycle. Different parts of your computer are out of sync with one another. If two parts are going to go back and forth quickly, circuit board designers need to place them physically close to one another, so they're not held back by the sluggish speed of light.

The problem that would really doom your VacPhone isn't speed. It's power. Vacuum tubes require a lot of electricity: 7AK7 vacuum tubes consume several watts while running, which means your phone would be putting out a total of 10^{11} watts worth of heat. How hot would it get? We can figure that out using the Stefan-Boltzmann law for radiated power:

PHONE SURFACE AREA · PHONE TEMPERATURE · ENVIRONMENT TEMPERATURE · PHYSICS STUFF

$$\text{POWER} = A \times \left(T_{\text{PHONE}}^{4} - T_{\text{ENVIRONMENT}}^{4}\right) \times e\sigma$$

$$T_{\text{PHONE}} = \sqrt[4]{\frac{\text{POWER}}{A_{\text{SURFACE}} \times e\sigma} - (T_{\text{ENV}})^4} = \sqrt[4]{\frac{10^{11}\text{ WATTS}}{100{,}000\text{M}^2 \times e\sigma} - (20^\circ\text{C})^4}$$

$$T_{\text{PHONE}} = 1{,}780^\circ\text{C}$$

Even if your phone were magically indestructible, the rest of the world isn't. A temperature of 1,780°C is above the melting point of granite, so if you dropped your phone, it might melt its way through the Earth's crust.

I recommend a protective case.

37. LASER UMBRELLA

Stopping rain from falling on something with an umbrella or a tent is boring. What if you tried to stop rain with a laser that targeted and vaporized each incoming droplet before it could come within ten feet of the ground?

—Zach

Stopping rain with a laser is one of those ideas that sounds totally reasonable, but if you—

While the idea of a laser umbrella might be appealing, it—

Okay. The idea of stopping rain with a laser is a thing we're currently talking about.

It's not a very practical idea.

First, let's look at the basic energy requirements. Vaporizing a liter of water takes about 2.6 megajoules,* and a big rainstorm might drop half an inch of rain per hour. This is one of those places where the equation isn't complicated—you just multiply the 2.6 megajoules per liter by the rainfall rate and you get the laser umbrella power requirement in watts per square meter of area covered. It's weird when units work out so straightforwardly:

$$2.6 \frac{\text{megajoules}}{\text{liter}} \times 0.5 \frac{\text{inches}}{\text{hour}} = 9,200 \frac{\text{watts}}{\text{meter}^2}$$

Nine kilowatts per square meter is an order of magnitude more power than is delivered to the surface by sunlight, so your surroundings are going to heat up pretty fast. In effect, you're creating a cloud of steam around yourself, into which you're pumping more and more laser energy.

In other words, you'd be building a human-size version of an autoclave, which is a piece of equipment used to sterilize objects through the incineration of organic matter within it. "Incineration of organic matter within it" is a bad feature for an umbrella.

But it gets worse! Vaporizing a droplet of water with a laser is more complicated than it sounds.† It takes a lot of energy—delivered fast—to vaporize the droplet without just splattering it apart into little droplets. Cleanly vaporizing a droplet would probably take more than the already unreasonable amounts we were considering.

* It takes more energy if the water is colder, but not *much* more. Heating the water up to the edge of boiling only takes a little of the 2.6 megajoules. Most of it goes into pushing it over the threshold from 100°C water to 100°C vapor.

† And to be honest, it sounds pretty complicated.

Then there's the problem of targeting. In theory, this is probably solvable. Adaptive optics technology, which is used to rapidly adjust telescope mirrors to cancel out the turbulence of the atmosphere, can allow for incredibly fast and precise control of beams of light. Covering an area of 100 square meters (which Zach also asked about in his full letter) would require something like 50,000 pulses per second. This is slow enough that you wouldn't run into any *direct* problems with relativity, but the device would—at minimum—need to be a lot more complicated than just a laser pointer on a swiveling base.

It might seem easier to forget about targeting completely and just fire lasers in random directions.* If you aim a laser beam in a random direction, how far will it go before it hits a drop? This is a pretty easy question to answer; it's the same as asking how far you can see in the rain, and the answer is at least several hundred meters. Unless you're trying to protect your whole neighborhood, firing powerful lasers in random directions probably won't help.

And, honestly, if you *are* trying to protect your whole neighborhood . . .

. . . firing powerful lasers in random directions *definitely* won't help.

* Really, what problem *can't* this strategy solve?

38. EAT A CLOUD

Could a person eat a whole cloud?

—Tak

No, unless you're allowed to squeeze the air out first.

Clouds are made of water, which is edible. Or drinkable, I guess. Potable? I've never been sure where the line between eating and drinking is.

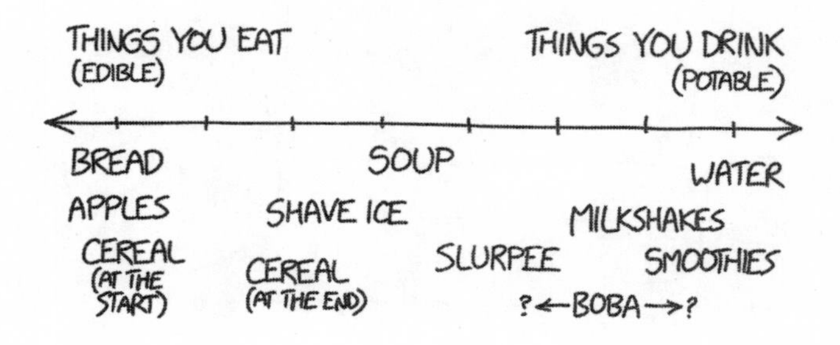

Clouds also contain air. We don't usually count air as part of food, since it escapes from your mouth as you chew or—in some cases—soon after you swallow.

You can certainly put a piece of a cloud in your mouth and swallow the water it contains. The problem is that you'll need to let the air escape—but air that's been inside your body will have absorbed a lot of moisture. When it leaves your mouth, it will carry that moisture with it, and once it encounters the cool, cloudy air, it will condense. In other words, if you try to eat a cloud, you'll just burp out more cloud faster than you can eat it.

But if you can collect the droplets together—perhaps by passing the cloud through a fine mesh and squeezing it out, or ionizing the droplets and collecting them on charged wires—you could absolutely eat a small cloud.

A fluffy cumulous cloud the size of a house contains about a liter of liquid water, or 2 or 3 large glasses, which is about the volume a human stomach can hold at one time. You couldn't eat a huge cloud, but you could absolutely eat one of those small house-size ones that briefly block the Sun for a second or two when they pass overhead.

A cloud is just about the largest thing you could eat in one sitting. There aren't a lot of things puffier and lower-density. Whipped cream seems pretty fluffy, but it's 15 percent as dense as water,* so a gallon of whipped cream would weigh about a pound. Even accounting for all the air that would escape, you couldn't eat more than a small bucket of it. Cotton candy, one of the most cloudlike foods, has a very low density—about 5 percent that of water—which means that you could in theory eat about a cubic foot of it in one sitting. That wouldn't necessarily be *healthy*, but it would be possible. But even if you spent your entire life eating cotton candy, you wouldn't be able to consume a house-size volume of it, especially since eating nothing but cotton candy would probably have an effect on your lifespan.

Other extremely lightweight edible substances include snow, meringues, and bags of potato chips, but the largest volume of each of them that you could eat in one sitting is about a cubic foot.

So if you want to eat a cloud, you'll need to do some work, but if you succeed, you'll have the satisfaction of knowing that you've eaten the largest thing you can possibly eat.

* Citation: Tracy V. Wilson, host of the podcast *Stuff You Missed in History Class*, who happened to have a cooking scale and a can of whipped cream on hand when I got this question.

A CLOUD
NUTRITION FACTS

SERVING SIZE: 1 CLOUD
SERVINGS PER SKY: COUNTLESS

TOTAL CALORIES: 0

	% DAILY VALUE*
TOTAL FAT: 0g	0%
SATURATED FAT: 0g	0%
TRANS FAT: 0g	0%
CHOLESTEROL: 0g	0%
SODIUM: 0g	0%
TOTAL CARBOHYDRATES: 0g	0%
DIETARY FIBER: 0g	0%
SUGAR: 0g	0%
PROTEIN:	SOMETIMES A FEW BUGS

CALCIUM: 0%	IRON: 0%*
MAGNESIUM: 0%	ZINC: 0%

*IRON VALUE MAY BE HIGHER IF YOU LIVE DOWNWIND FROM THE HOUSE FROM CHAPTER 4

Just remember to store your cloud in a reusable bottle. There's no need to waste all that plastic!

39. TALL SUNSETS

Let's say that two people of different heights (159 cm and 206 cm) stand beside each other while looking at the sunset. How much longer will the taller person be able to see the Sun than the shorter one?

—Rasmus Bunde Nielsen

Over a full second longer!

The Sun sets later for taller people, because the higher up you are, the farther you can see over the horizon.

In addition to later sunsets, taller people also experience earlier sunrises, which means that days last longer for them in general. If you're near the equator at sea level, every extra inch of height corresponds to nearly a minute of extra daylight per year, and it's even more at higher latitudes. At 100 feet above sea level, the effect is smaller, but each inch of height still gains you at least an extra 10 seconds of daylight annually.

On the other hand, tall people experience stronger winds, hit their heads more often while going up stairs, walk into more spiderwebs, and are more likely to be decapitated by swinging blades after accidentally wandering into a booby-trapped ancient temple. (I don't know exactly what the probability of *that* happening is, but I know it must increase with height.)

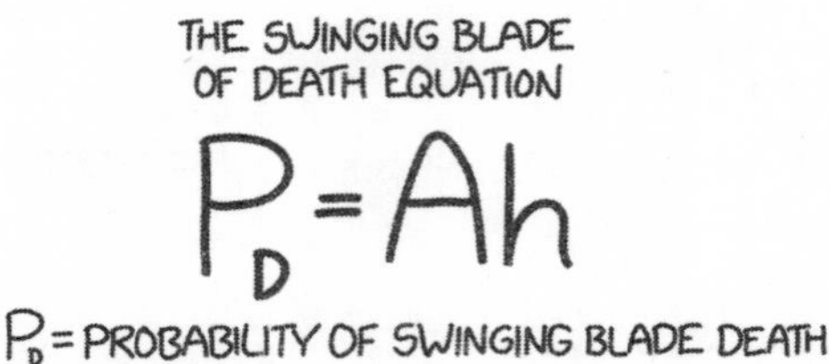

If you have a good view of the horizon near sea level, you can use this height effect to see two sunrises or sunsets in a row. All you need is a staircase, ladder, or hill that you can move up or down quickly.

It's easier to do this with the sunrise than the sunset, since going up the stairs fast is harder than going down, but it does mean waking up early.

On the other hand, if getting more sunlight is your goal, then waking up early might be its own reward. If you live around sea level and you normally sleep late, then getting up 10 seconds earlier each day will let you experience extra daylight, which is equivalent to adding 20 feet to your height.

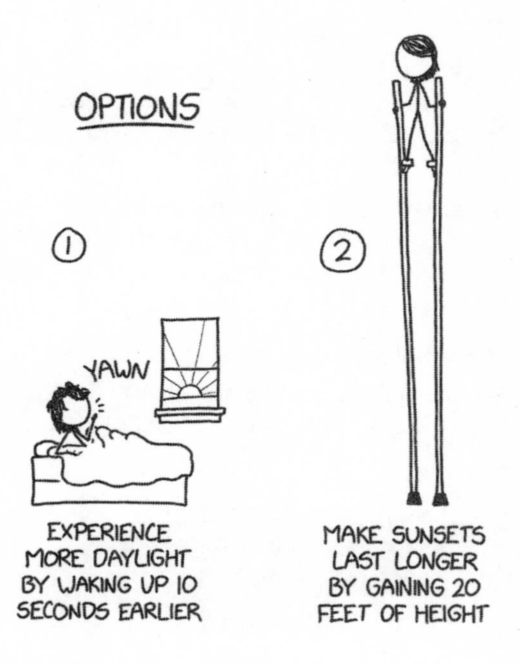

Still, it's nice to sleep in.

40. LAVA LAMP

What if I made a lava lamp out of real lava? What could I use as a clear medium? How close could I stand to watch it?

—Kathy Johnstone, sixth-grade teacher (via a student)

This is a surprisingly reasonable idea, by *What If* standards.

I mean, it's not *that* reasonable. At the very least, I'm guessing you would lose your teaching license and possibly some of the students in the front row. But you could do it.

WHY DID YOU SPRAY YOUR STUDENTS WITH STEAM, BROKEN GLASS, AND MOLTEN ROCK?
IF WE ARE TO TEACH THEM, THEY MUST FEAR US.

You have a few choices for transparent materials that could hold the lava without rupturing and splattering half the classroom with red-hot droplets. Fused quartz glass would be a great choice. It's the same stuff they use in high-intensity lamp bulbs, the surface of which can easily get up to midrange lava temperatures.[*] Another possibility is sapphire, which stays solid up to 2,000°C and is commonly used as a window into high-temperature chambers.

The question of what to use for the clear medium is trickier. Let's say we find a transparent glass that melts at low temperatures. Even if we ignore the impurities from the hot lava that would probably cloud the glass, we're going to have a problem.[†]

Molten glass is transparent. So why doesn't it *look* transparent?[‡] The answer is simple: It glows. Hot objects give off blackbody radiation; molten glass glows just like molten lava does, and for the same reason.

So the problem with a lava lamp is that both halves of it will be equally bright, and it will be hard to see the lava. We could try having nothing in the top half of the lamp—after all, when it's hot enough, lava bubbles pretty well on its own. Unfortunately, the lamp itself would *also* be in contact with the lava. Sapphire might not melt easily, but it *will* glow, making it hard to see whatever the lava inside is doing.

---

* Some bulbs for stage lights advertise that they can handle temperatures of up to 1,000°C, which is hotter than many types of lava.

† And later, when the school board finds out about this, we'll have another.

‡ Which sounds sorta contradictory. "This music is loud, but it doesn't *sound* loud."

Unless you hooked it up to a really bright bulb, this lava lamp would cool down quickly. Just like individual blobs of lava dropped on the ground in real life, the lamp would solidify and stop glowing within the first minute, and by the end of the class period you'd probably be able to touch it without being burned.

A solidified lava lamp is just about the most boring thing in the world. But the scenario made me wonder: If making a lamp out of molten lava wouldn't be very exciting, then what about a volcano made of lamps?

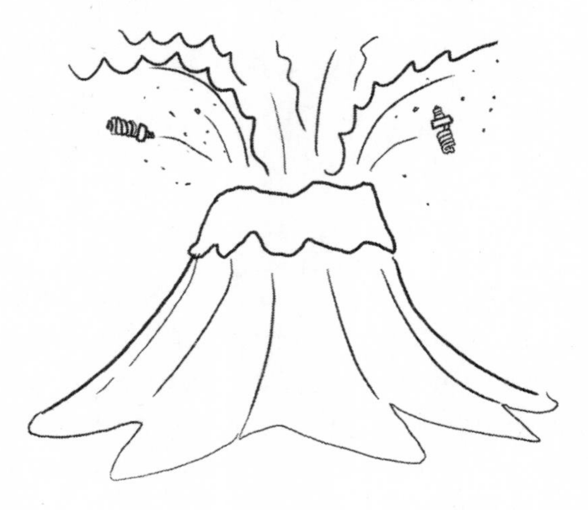

This is probably the most useless calculation I've ever done,* but . . . what if Mount Saint Helens erupted again today, but instead of tephra,† it spewed compact fluorescent bulbs?

Well, if it did, the mercury released into the atmosphere would be several orders of magnitude larger than all man-made emissions combined.‡

* Okay, there's no *way* that's true.

† The technical term for "whatever that stuff is that comes out of a volcano."

‡ 45 percent of which come from gold mining.

I like how it's totally not clear what the rest of this claim is supposed to be. "THE MORE YOU KNOW . . ." . . . what? The happier you are? The more cultured you are? Are you better able to survive a life-or-death trivia contest? If I were doing the show I would replace it with "YOU JUST LEARNED THAT."

All in all, I think making a lava lamp out of lava would be kind of anticlimactic. I also think that it's probably good that Mount Saint Helens didn't erupt compact fluorescent bulbs. And I think that if I were in Ms. Johnstone's class, I'd try to sit toward the back of the room.

41. SISYPHEAN REFRIGERATORS

Suppose everyone with a fridge or a freezer opened them at the same time, outdoors. Would that amount of cooling be able to noticeably change the temperature? If not, how many fridges would it take to lower the temperature, say, 5 degrees F? What about even lower?

—Nicholas Mittica

Refrigerators don't cool their surroundings, they heat them.

Refrigerators work by pumping heat from their interior to their exterior. The inside gets colder, and the outside gets hotter. If you open the door, the fridge will struggle endlessly to draw up heat from the front and disperse it out into the air via the coils, only to have the air flow right back in. Then it has to start all over, like Sisyphus forever rolling a boulder up a hill.

In order to move all this heat around, the refrigerator consumes electricity, which produces additional heat. A refrigerator running its compressor at full power, as it would if you left the door open, might consume 150 watts. That means that on top of the heat that it would pointlessly transfer from the interior to the coils in the back, an additional 150 watts worth of heat would be dumped into the surrounding environment.

That extra 150 watts of heat per refrigerator would technically raise the average temperature of the Earth, but only a little. There are probably a few hundred million homes with refrigerators right now, but even if we assume that every one of the 8 billion people in the world owned a refrigerator, and they all left them running outside 24/7, the global temperature increase would be less than 1/1,000th of a degree Celsius, which isn't nearly enough to measure.

But even though their direct waste heat would be negligible, those refrigerators *would* make the Earth hotter. A lot of the electricity in our homes comes from burning fossil fuels. If those 8 billion outdoor refrigerators were powered by a mix of power sources similar to that of the United States in 2022, they would add about 6 billion tons of CO_2 to the atmosphere each year, about 15 percent of global emissions.

If the refrigerators kept up that emission at the same rate for the rest of the twenty-first century, climate models suggest it would add an extra 0.3°C of global warming on top of whatever other warming humans cause.

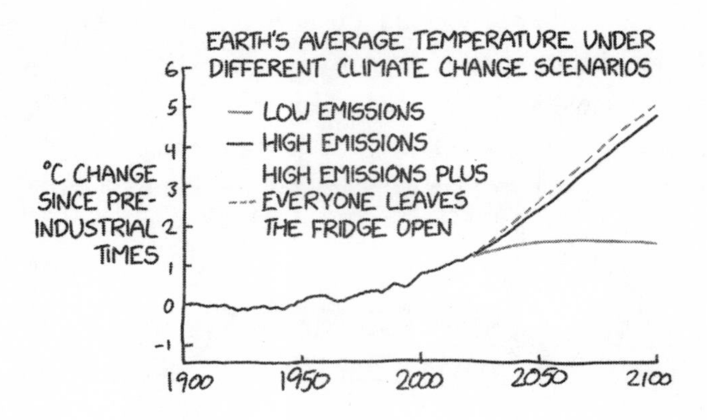

How does this compare to other pointless tasks? Greek mythology tells of Sisyphus rolling a boulder up a hill forever. Homer's description in *The Odyssey* makes it clear he's working pretty hard:

> *And I saw Sisyphus at his endless task raising his prodigious stone with both his hands. With hands and feet he tried to roll it up to the top of the hill, but always, just before he could roll it over on to the other side, its weight would be too much for him, and the pitiless stone would come thundering down again on to the plain. Then he would begin trying to push it up hill again, and the sweat ran off him and the steam rose after him.*

—*The Odyssey*, Samuel Butler translation, 1900

Data from ultramarathoners shows that the limit on the amount of work humans can do during long-term endurance events is 2.5 times their resting metabolic rate. I have no idea how to even begin to come up with a reasonable estimate for Sisyphus's caloric intake, but he clearly works out a lot, so let's use famously buff wrestler/actor Dwayne Johnson as a stand-in. I looked up Johnson's height and weight and plugged them into a resting metabolism calculator, which gave an estimate of 2,150 calories/day, or 105 watts.

	SISYPHUS OF CORINTH	DWAYNE JOHNSON
VERY STRONG	YES	YES
FAMOUS FOR TALES IN WHICH HE REPEATEDLY ESCAPED DEATH	YES (THE MYTH OF SISYPHUS)	YES (FAST & FURIOUS FRANCHISE)
ONCE TOOK THE PLACE OF A GOD	YES (THANATOS, IN TARTARUS)	YES (MAUI, IN MOANA)
HEIGHT AND WEIGHT EASY TO GOOGLE	NO	YES

Using 105 watts for Sisyphus's metabolic rate, we can estimate that his maximum long-term output would be 260 watts, or a little more than an open refrigerator.

So if you want to have a pointless object in your front yard wasting energy forever for no good reason, then instead of plugging in your refrigerator, just have Sisyphus push a rock up a hill. It would reduce your electric bill, and the climate-change impact would be negligible, since the power would come from a renewable energy source (the infinite spite of Hades, God of the Underworld).

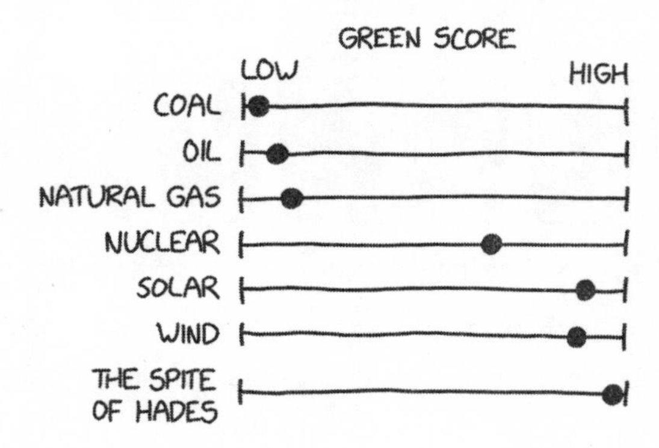

If Sisyphus is unavailable, maybe you can get Dwayne Johnson to help out instead.

DWAYNE "THE ROCK" JOHNSON

42. BLOOD ALCOHOL

Could you get drunk from drinking a drunk person's blood?

—Fin Byrne

You would have to drink a lot of blood.

A person contains about 5 liters of blood, or 14 glasses.

Remember, you're supposed to drink 8 glasses of blood per day.

If your blood is more than about half a percent alcohol, you stand a pretty good chance of dying. There have been a handful of cases of people surviving with a blood alcohol level of above 1 percent, but the LD_{50}—the level at which 50 percent of people will die—is 0.40 (0.4 percent).

If someone had a blood alcohol concentration (BAC) of 0.40, and you drank all 14 glasses of their blood in a short amount of time,[*] you would throw up.

I DON'T WANT TO DRAW SOMEONE
THROWING UP A GALLON OF BLOOD

SO HERE'S A SQUIRREL INSTEAD.

You wouldn't throw up because of the alcohol; you'd just throw up because you're drinking blood. If you somehow avoided vomiting, you would have ingested a total of 20 grams of ethanol, which is the amount you'd get from a pint of beer.

Depending on your weight, drinking that much blood could raise your own blood alcohol level to about 0.05. This is low enough that you could legally drive in many jurisdictions, but high enough to double your risk of an accident if you did.

If your BAC is 0.05, it means only ⅛th of the alcohol from the other person's blood made it into yours. If, after you drank all this blood, someone killed you and drank *your* blood,[†] they would then have a BAC of 0.006. If this process were repeated about 25 times, there would be fewer than 8 molecules of ethanol left in the last person's blood. After a few more cycles, there would likely be none;[‡] they'd just be drinking regular blood.[§]

* If you drink all of someone's blood, there's a 100 percent chance that they'll die.

† It's only fair.

‡ By homeopathic standards, this is still quite concentrated.

§ Like a loser.

Whether there's any alcohol in it or not, drinking 14 glasses of blood wouldn't be fun. There's not a huge amount of medical literature on the subject, but anecdotal evidence from some particularly alarming internet forum posts suggests that any normal person who tries to drink more than about a pint of blood will vomit, as you can see from this illustration:

If you drink blood regularly, over a long period of time the buildup of iron in your system can cause iron overload. This syndrome, which sometimes affects people who have repeated blood transfusions, is one of the few conditions for which the correct treatment is bloodletting.

Drinking one person's blood probably wouldn't cause iron overload. What it *could* give you is a blood-borne disease. Most such diseases are caused by viruses that can't survive in the stomach, but they could easily get into your blood through scratches in your mouth or throat as you drink.

Diseases you could get from drinking an infected person's blood include hepatitis B and C, HIV, and viral hemorrhagic fevers such as hantavirus and Ebola. I'm not a doctor, and I try not to give medical advice in my books. However, I will confidently say that you shouldn't drink the blood of someone with a viral hemorrhagic fever.

THINGS YOU SHOULD NOT DO
(UPDATED LIST)

#156,818 PEEL AWAY THE EARTH'S CRUST
#156,819 TRY TO PAINT THE SAHARA DESERT BY HAND
#156,820 REMOVE SOMEONE'S BONES WITHOUT ASKING
#156,821 SPEND 100% OF YOUR GOVERNMENT'S BUDGET ON MOBILE GAME IN-APP PURCHASES
#156,822 FILL A LAVA LAMP WITH ACTUAL LAVA
#156,823 (NEW!) DRINK THE BLOOD OF SOMEONE WITH A VIRAL HEMORRAGHIC FEVER

That said, drinking or eating blood is not unheard of. It's a taboo in many cultures, but "black pudding," which is largely blood, is a traditional British dish, and there are similar dishes all around the world. Maasai pastoralists in east Africa once lived mainly on milk, but also sometimes drank blood, drawing it from their cattle and mixing it with the milk to form a sort of extreme protein shake.

So the bottom line is that drinking enough of someone's blood to get drunk would be very difficult, probably quite unpleasant, and might give you some serious diseases. It wouldn't matter how drunk they were—the blood itself would do awful things to your body long before the booze ever could.

43. BASKETBALL EARTH

You know how when you spin a basketball on your finger you hit the side to make it go faster and balance it? If a meteor passes close enough to the Earth, can it make the Earth spin faster like your hand does the basketball?

—Zayne Freshley

Yes!

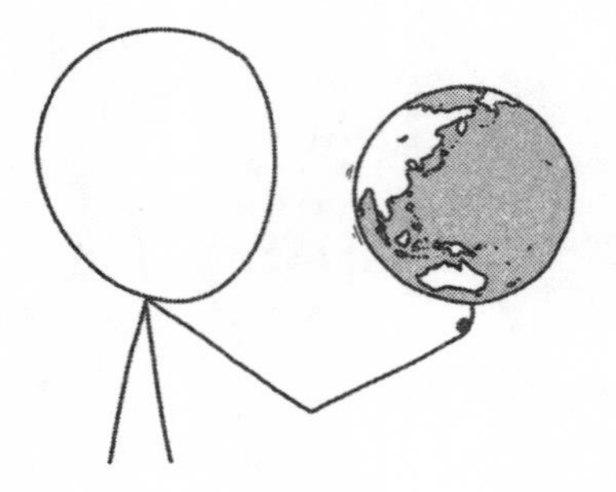

This is one of those things that seems like it shouldn't work that way, but it turns out it works *exactly* that way.

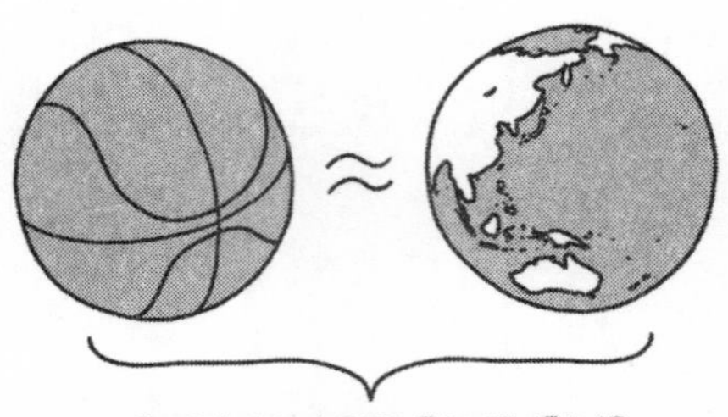

When meteors hit the Earth, or skim through the atmosphere, they alter the planet's spin.

Meteors aren't usually going perfectly straight down when they enter the atmosphere. Unless they happen to be aimed exactly right, they hit at an angle, and so they give the Earth a little spin in one direction or the other. If they're going east, they speed the planet up, and if they're going west, they slow it down.

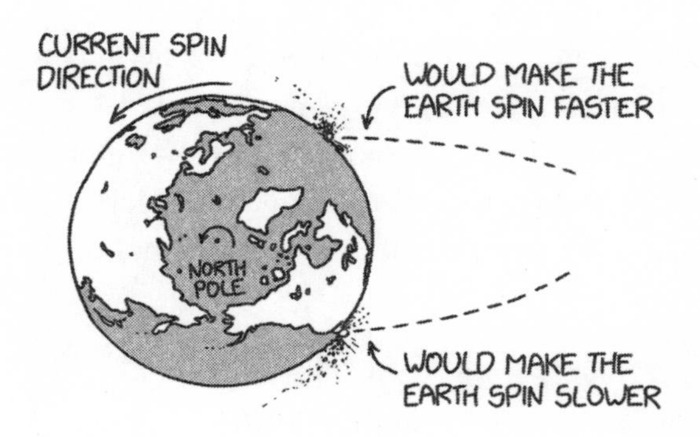

A meteor that just flies past the Earth through space doesn't measurably affect its rotation; it has to make physical contact with the planet. But it doesn't actually have to reach the ground. If it burns up in the atmosphere, its debris still gives the air a big push, and some of that moving air eventually pulls on the ground through drag.

Even if the meteor skims the atmosphere and then returns to space, much of the momentum it loses in the atmosphere ultimately gets transferred to the Earth's rotation. These Earth-grazing fireballs are rare, but one glanced off the atmosphere over the western United States and Canada in 1972, and others have been spotted by sky-watchers, automated telescopes, and radar.

The Earth is big,[citation needed] so even devastating meteor strikes aren't likely to change the length of the day that much. The dinosaur-killing Chicxulub impact, which left a crater 100 kilometers across, probably only changed the length of the day by a few milliseconds at most. For most purposes, a few milliseconds of change isn't enough to notice, although it would mean they'd need to add a leap second every year to account for it.

If something comparable in size to a moon or planet hits us, it *could* drastically change the length of the day, at the cost of much greater destruction. We think the Moon was probably created from debris when a Mars-size object hit the Earth as it formed. That impact probably made a big change to the length of the day. In a sense, it also made an even *bigger* change to the length of the month . . .

THE FIRST EVER MONTHLY CALENDAR

. . . by creating months in the first place.

44. SPIDERS VS. THE SUN

Which has a greater gravitational pull on me: the Sun or spiders? Granted, the Sun is much bigger, but it is also much farther away, and as I learned in high school physics, the gravitational force is proportional to the square of the distance.

—Marina Fleming

In the literal sense, this question is totally reasonable, although it would be easy to rephrase it to be completely incoherent.

The gravitational pull from a *single* spider, no matter how heavy, will never beat out the Sun. The goliath bird spider[*] weighs as much as a large apple.[†] Even if, God forbid, you were as close as possible to one of them, the pull from the Sun would still be 50 million times stronger.

What about *all* the spiders in the world?

There's a well-known factoid that claims you're always within a few feet of a spider. This isn't literally true—spiders don't live in the water,[‡] so you can get away from them by swimming, and there aren't as many spiders in buildings as in fields and forests. But if you're anywhere near the outdoors, even in the Arctic tundra, there are probably spiders within a few feet of you.

Regardless of whether the factoid is precisely true or not, there are an awful lot of spiders out there. Exactly how many is hard to say, but we can do some rough estimation. A 2009 study of spider density in Brazil found one-digit numbers of milligrams of spider per square meter of forest floor.[§] If we guess that about 10 percent of the world's land area hosts this density of spiders, and there are none anywhere else, we come up with 200 million kilograms worldwide.[¶]

Even if our numbers are wildly off, it's enough to answer Marina's question. If we assume the spiders are distributed evenly across the surface of the Earth, we can use Newton's shell theorem to determine their collective gravitational pull on objects outside the Earth. If you do that math, you find that the Sun's pull is stronger by *13 orders of magnitude*.

Now, this calculation makes some assumptions that aren't true. Spider distributions are discrete, not

* Wikipedia helpfully notes that, despite its name, it "only rarely preys on birds."

† This is correct whether I mean the fruit or an iPhone; the spider weighs about as much as each.

‡ With the exception of *Argyroneta aquatica*, the diving bell spider.

§ That's dry mass; you have to multiply by 3 or 4 to get the live weight.

¶ One survey of fields and pastures in New Zealand and England tended to find two-digit numbers of spiders per square meter. If they each weigh about a milligram, and we assume once again that about 10 percent of Earth's land supports that density of spiders, that gives a total spider biomass of 100 million to a billion kilograms. That agrees with our first estimate, at least.

continuous,* and some areas have more spiders than others. What if there happen to be a *lot* of spiders near you?

In 2009, the Back River Wastewater Treatment Plant found themselves dealing with what they called an "extreme spider situation." As described in a fascinating and horrifying article published by the Entomological Society of America,† an estimated 80 million orb-weaving spiders had colonized the plant, covering every surface with heavy sheets of web.‡

What was the total force of gravity from all those spiders? First we need their mass; according to a paper titled "Sexual Cannibalism in Orb-Weaving Spiders: An Economic Model,"§ it's about 20 grams for males and several times that for females. So even if you were standing next to the Back River Wastewater Treatment Plant in 2009, the pull of all the spiders inside would still be only 1/50,000,000th that of the Sun.

No matter which way you look at it, the bottom line is that we live our lives surrounded by tiny spiders on a world completely dominated by a gigantic star.

Hey, at least it's not the other way around.

* Spiders are quantized.

† The conclusion of the article contains this absolutely incredible passage:
Our recommendations for amelioration included the following general points:
1) On-site personnel should be reassured that the spiders are harmless and the facility's immense shroud of silk should be presented in a positive light as a record-breaking natural history wonder.

‡ Which was in turn covered in heavy sheets of spider.

§ Not to be confused with "Trade-off between pre- and postcopulatory sexual cannibalism in a wolf spider," which is a different but equally real paper.

45. INHALE A PERSON

If house dust comprises up to 80 percent dead skin, how many people worth of skin does a person consume/inhale in a lifetime?

—Greg, Cape Town, South Africa

Good news: You can't inhale a person, and also dust is not mostly dead skin.

The claim that household dust is mostly dead skin is widespread; if you google it, you'll find a lot of articles both supporting and debunking it.* Part of why this is hard to pin down is that household dust isn't any one specific thing. It's just a disgusting salad made from whatever happens to be lying around your house. It can include soil,

* Derek Muller of the YouTube channel Veritasium did a lengthy video on this question, citing a 1981 book that in turn cited a 1967 Dutch cleaning standards publication. He ultimately came down on the side of "Wow, there's a lot of skin there."

pollen, cotton fibers, crumbs, powdered sugar, glitter, pet hair and dander, plastic, soot, human or animal hair, flour, glass, smoke, mites, and countless blobs of hard-to-identify gunk stuck together.

There's definitely some skin in there, but it's not usually the main ingredient. Surveys of dust from the floors of offices and schools have found that a majority of it wasn't organic matter at all, and a 1973 study in *Nature* of various environments found that skin cells made up between 0.4 percent and 10 percent of the airborne dust.

We do spew out dead skin at a ridiculous rate. We shed something like 50 milligrams of cells per hour, but most of that skin doesn't go into the air. If we were pushing 50 milligrams of skin dust into the air every hour, our houses would be as dusty as coal mines or wood shops. Since the air isn't constantly full of dust, it must be going somewhere else. Some of it settles quickly onto the floor, but a lot of it goes down the drain when we wash, rubs off on our clothes and gets washed away by detergent, or ends up on our pillows and mattresses.

Even if you found a way to maximize the airborne-skin dust concentration, you wouldn't be able to inhale a person. If you built a machine to pump skin dust into a room, and you raised the concentration to 10 mg/m^3—making the air so dusty that it would exceed the occupational dust exposure limits for coal mine workers—you would still only inhale about 3 kilograms of skin cells over the average lifetime.

So no, you can't inhale a person, but you *can* inhale a larger fraction of a person than I think anyone is really comfortable with.

Also, I don't think I want to answer any more questions about skin.

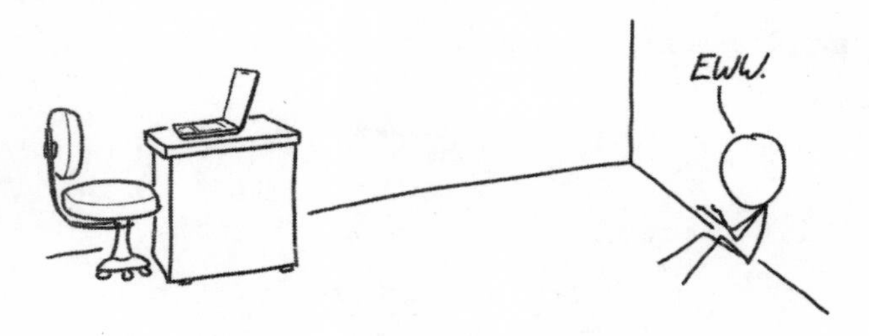

46. CANDY CRUSH LIGHTNING

How many Wint-O-Green Life Savers would it take to create a life-size lightning bolt if you crushed them?

—Violet M.

Billions.

When you crush sugar in the dark, it emits flashes of light. This phenomenon is called triboluminescence. The light can be pretty faint, but the old Wint-O-Green flavor* of Life Savers candies are famous for producing an especially bright flash, which is thanks to an additive used for flavoring. Most of the light emitted by sugar through triboluminescence is ultraviolet, but certain Life Savers contain methyl salicylate, which is fluorescent. It absorbs the invisible ultraviolet and emits it as blue visible light.

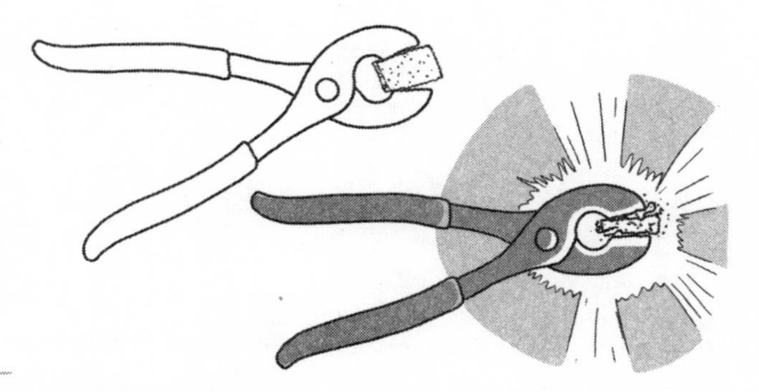

* Which has apparently been spelled like that all along and I never noticed until now. I guess the *O* in Wint-O-Green is like the *a* in Berenstain Bears.

We don't really understand triboluminescence.

When materials scrape together or are split into pieces, electric charges are sometimes pulled apart in a way that lets them snap together to release energy. But there are a lot of ways atoms can bonk against one another, and scientists have trouble figuring out exactly what combination of effects is producing light in any particular experiment.

If you bite down on a Life Saver with a force of 20 pounds in order to crush it, you deliver about a joule of mechanical energy into the sugar crystals.* By comparison, a lightning strike carries about 5 or 10 billion joules of energy, so to get the same amount of energy to work with, you'd need to crunch 5 or 10 billion Life Savers.

* Some triboluminescence might involve the release of stored chemical energy, which could reduce the number of Life Savers required for a given flash.

Crushing a Life Saver doesn't really produce a spark. The spark when you touch a doorknob really is a spark; if you look at it up close, it looks like a tiny bolt of lightning. But if you look closely at slow-motion photography of Life Savers breaking, you won't see a lightning bolt. The sugar just glows briefly as it breaks, like a flashbulb firing. But despite their different appearances, Life Savers flashes and lightning have a lot in common. They both involve electric charges being pulled apart by materials mechanically rubbing against one another, and in both cases light is produced by the energy release when those charges equalize.

And when it comes down to it, we don't understand lightning, either. We know updrafts in storms cause electric charges to build up between the top and bottom of the storm, and we think it involves the wind blowing past rain or ice, but the details of how the charges separate are still a mystery.

short answers #4

Q Can humans safely eat rabid creatures?

—Winston

No. Eating an animal with rabies is not safe and may transmit rabies. There are several cases in the medical literature of patients with rabies who are believed to have caught the virus by eating infected animals.

WHAT YOU'D EXPECT THE ANSWER TO BE

WHAT THE ANSWER ACTUALLY IS	YES	NO
YES	DOES MIT HAVE CLASSROOMS?	DOES AMHERST COLLEGE HAVE A NUCLEAR BUNKER?
NO	DO SCIENTISTS KNOW WHY LIGHTNING HAPPENS?	IS IT SAFE TO EAT RABID ANIMALS?

What if the Earth's core suddenly stopped producing heat?

—Laura

Honestly, we'd be fine.

Any instantaneous physical change in the Earth could in theory change the stress within the crust and cause earthquakes and volcanic eruptions, but if you assume whatever caused the core to stop producing heat also gently redistributed those short-term stresses, then the actual change in heat flow wouldn't really be a problem.

Most of our heat comes from the Sun. The flow through the crust is such a small part of the Earth's overall surface heat balance that it wouldn't affect the atmosphere much. If the outer core solidified, we'd lose our magnetic field, but—despite what the 2003 film *The Core* will tell you—that wouldn't cause microwave beams from space to cut the Golden Gate Bridge in half or anything. It would just slightly increase the rate at which our upper atmosphere is lost to space.

Over a long enough time, plate tectonics—which is powered by the Earth's internal heat—would grind to a halt. Plate tectonics are a key part of the long-term carbon cycle, which regulates the Earth's temperature, so eventually that thermostat would fail and the oceans would boil away. But that's going to happen anyway, so I wouldn't worry about it.

Q Could humanity, with our current technology, destroy the Moon?

—Tyler

Q Can global warming cause the Earth's magnetic fields to weaken?

—Pavaki

Q If you used a laser, would you be able to bake something?

—Andrew Liu

No, no, and yes, respectively.

CAN WE... / USING...

	LASERS	ALL HUMAN TECHNOLOGY	GLOBAL WARMING
DESTROY THE MOON	NO	NO	NO
WEAKEN EARTH'S MAGNETIC FIELD	NO	NO	NO
BAKE COOKIES	YES	YES	IF IT GETS REALLY BAD

Q What if Earth was sliced in half, like an apple? Where should you be such that you have the best chance of survival?

—Anonymous

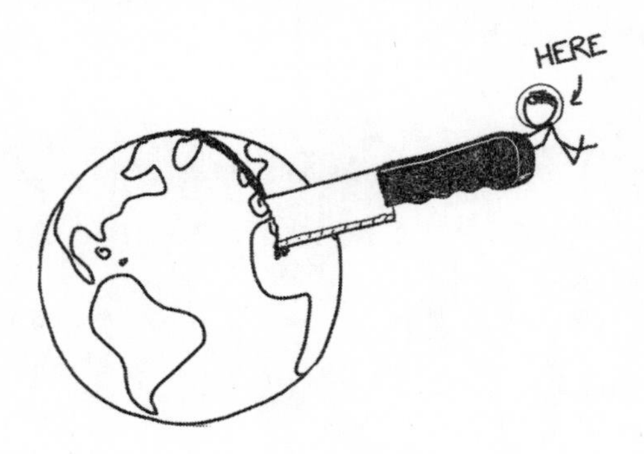

Q What would happen if a person dropped into a pool full of jellyfish?

—Lorenzo Belotti

It depends on the species. The largest group of jellyfish I've ever seen were moon jellies, whose sting is often so mild that humans don't even notice it. They feel surprisingly firm to the touch, like wet gummy candies. So it's possible the person would just make some slippery new friends!

YOU'RE MY NEW BEST FRIEND. I'M GOING TO CALL YOU JELLY DONUT!

WHEN THE SEA OVERTAKES YOUR CITIES, MY KIND WILL FLOAT THROUGH YOUR STREETS AND FEED UPON YOUR RUINS.

AWW, SO CUTE!

Q Would it be possible to make a house floor into a massive air hockey table, so you could move heavy furniture across the room?

—Jacob Wood

Yes, and I know what my next home-improvement project is going to be.

Q My 7-year-old son asked us over dinner recently at which point potatoes melt (I assume in a vacuum). Please advise.

—Steffen

Potatoes don't really melt at any temperature. The starches break down and gelatinize, which is part of the normal cooking process; as the heat rises, the different components will sublimate at different temperatures.

But what I want to know is, do you normally add "in a vacuum" to all his questions and assume that's what he meant?

CAN I HAVE A PIZZA PARTY FOR MY BIRTHDAY?
YOU WANT A PIZZA PARTY IN A VACUUM? THAT WILL BE TOUGH, BUT WE CAN TRY...

> **Q** Would a pigeon be able to make it to space if it was not affected by gravity?
>
> **—Nick Evans**

No. Birds can flap around in zero gravity and might be able to propel themselves along, but it's too cold in the upper atmosphere and pigeons need to breathe.

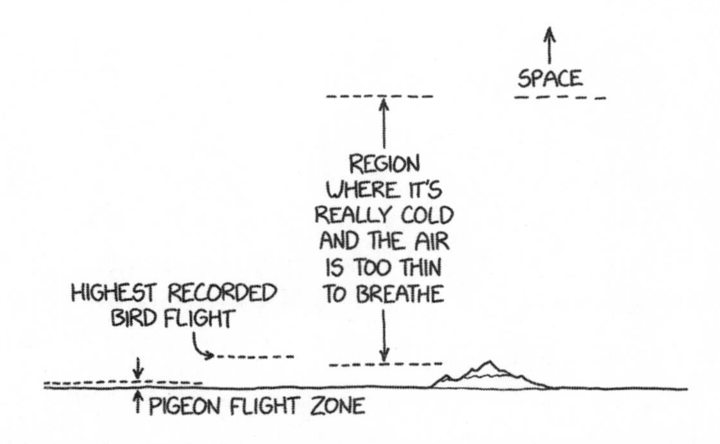

> **Q** If you were flying blind through the Milky Way, what would be the odds of hitting a star or planet?
>
> **—David**

Even if you flew through edge-on, so you spent as much time as possible in the dense galactic disk, your odds of hitting a star would be only about 1 in 10 billion. (Your odds of hitting a planet would be a thousand times smaller.)

For comparison, that's about the same as the odds of deciding to call Barack Obama, picking up your phone and dialing 10 random digits, and getting his cell number on your first try.

The flight across the galaxy would take a long time, though. If you try a number every 30 seconds, it would only take you 10,000 years to dial them all. The trip across the galaxy will take a lot longer—10 million years at 1 percent of the speed of light—so that will give you and Obama plenty of time to chat once you get his number.

Q On various bodies in our Solar System (feel free to group any that are equivalent), roughly how long could you typically survive on the surface (for gas giants, assume you are on a magical platform at some point in the atmosphere that you could reasonably treat as the surface) with nothing but an infinite air supply and warm winter clothing? That is, no helmet, no pressure suit, just a nose-and-mouth air mask attached to a magic air generator, and clothing that would be suitable for, say, Chicago in winter. (No cute tricks like using the magic air supply to generate heat or whatever.)

—Melissa Trible

- **Earth:** 100-ish years
- **Venus:** Weeks to months
- **Everywhere else:** Minutes to hours

There's a layer in the atmosphere of Venus where the temperature and pressure are both relatively close to normal Earth surface conditions—the only place in the Solar System like that other than Earth and the interiors of spacecraft. But I imagine the sulfuric acid fog on your skin would get a little old before too long.

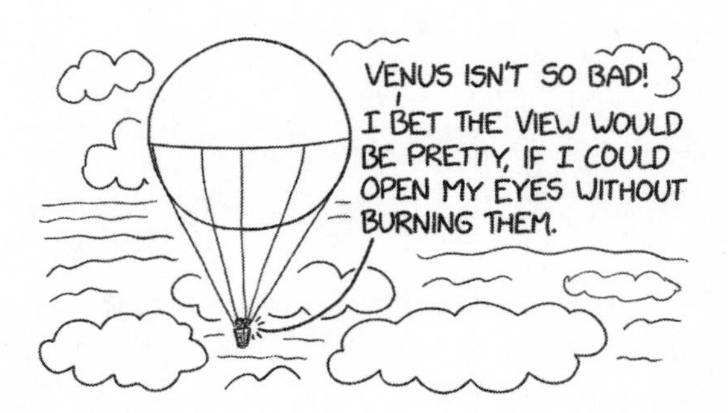

Q What would happen if someone dropped an anvil on you from space?

—Sam Stiehl, age 10, Evanston, IL

The good news is that an anvil is small enough that the atmosphere would slow it down to terminal velocity by the time it reached you. The bad news is that the terminal velocity of an anvil is roughly 500 miles per hour.

When an anvil lands on you, it doesn't really matter how high it fell from.

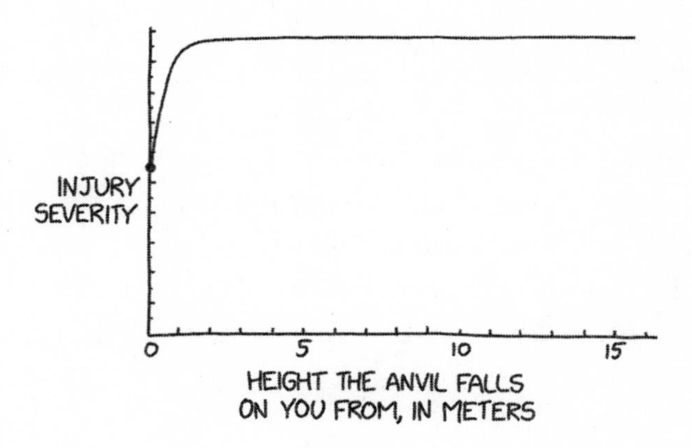

47. TOASTY WARM

What if I want to heat my house using toasters. How many do I need?

—Peter Ahlström, Sweden

Not very many, since your house will probably catch fire if you leave toasters running all the time. Once it does, the house will become self-heating for the rest of its lifetime.

I'VE DISCOVERED A WAY TO MAKE OUR HOUSE SELF-HEATING FOR ABOUT 15 OR 20 MINUTES!

But for the short time before your house caught fire, toasters would keep it warm just fine.

Electric space heaters aren't always the best way to heat a home—using electricity to directly produce heat is generally less efficient than using that power to warm up outside air using a heat pump, and in some places electricity can also be more expensive than natural gas or oil heat. But one neat thing about space heaters is that they're all equally efficient: All space heaters produce one watt of heat for each watt of electricity they draw.

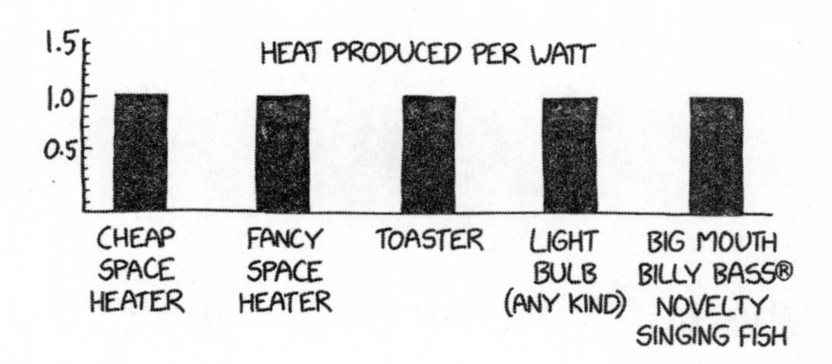

In fact, thanks to the laws of thermodynamics, just about *every* electric device that consumes power eventually turns that power into heat at the same rate. A 60-watt lightbulb produces light, but that light hits a surface and heats it up. In the end, it produces the same 60 watts of heat as a 60-watt space heater. Toasters, blenders, microwaves, and lightbulbs all produce heat at the rate of 1 watt per watt, just like a space heater.

An average toaster uses about 1,200 watts of electric power, and a heating system for a typical house in the northern United States might need to supply 80,000 BTUs/hour, which works out to 25,000 watt-hours per hour, or 25,000 watts. Heating one of these houses would take about 20 toasters.

If you don't want to run your toasters empty, you could try making lots of toast, but you'll quickly have more than you can eat. If each toaster can hold two slices, and it takes about 2 minutes to toast each one, then your toaster will go through about 30 loaves of bread per hour. At peak, you'll be consuming bread at the rate of a medium-size American town.

THIS IS THE WORST IDEA FOR HEATING A HOUSE SINCE, OR WITH, SLICED BREAD.

48. PROTON EARTH, ELECTRON MOON

What if the Earth were made entirely of protons, and the Moon were made entirely of electrons?

—Noah Williams

This might be the most destructive *What If* scenario I've written about.

You might imagine an electron Moon orbiting a proton Earth, sort of like a gigantic hydrogen atom. On one level, it makes a kind of sense; after all, electrons orbit protons, and moons orbit planets. In fact, a planetary model of the atom was briefly popular (although it turned out not to be very useful for understanding atoms*).

* This model was largely obsolete by the 1920s, but it lived on in an elaborate foam-and-pipe-cleaner diorama I made in sixth-grade science class at Salem Church Middle School.

If you put two electrons together, they try to fly apart. Electrons are negatively charged, and the force of repulsion from this charge is about 20 orders of magnitude stronger than the force of gravity pulling them together.

If you put 10^{52} electrons together—to build a moon—they would push one another apart so hard that each electron would be shoved away with an *unbelievable* amount of energy.

It turns out that, for the proton Earth and electron Moon in Noah's scenario, the planetary model is even more wrong than usual. The Moon wouldn't orbit the Earth because they'd barely have a chance to influence each other; the forces trying to blow each one apart would be far more powerful than any attractive force between the two.

If we ignore general relativity for a moment—we'll come back to it—we can calculate that the energy from these electrons all pushing on one another would be enough to accelerate all of them outward at near the speed of light.* Accelerating particles to those speeds isn't unusual; a desktop particle accelerator—for example, a CRT monitor—can accelerate electrons to a reasonable fraction of the speed of light. But the electrons in Noah's Moon would each be carrying much, much more energy than those in a normal accelerator. Their energy would be orders of magnitude more than the Planck energy, which is itself many orders of magnitude larger than the energies we can reach in our largest accelerators. In other words, Noah's question takes us pretty far outside normal physics, into the highly theoretical realm of things like quantum gravity and string theory.

So I contacted Dr. Cindy Keeler, a string theorist with the Niels Bohr Institute, and asked her about Noah's scenario.

* But not past it; we're ignoring general relativity, but not special relativity.

Dr. Keeler agreed that we shouldn't rely on any calculations that involve putting that much energy in each electron since it's so far beyond what we're able to test in our accelerators. "I don't trust anything with energy per particle over the Planck scale," she said. "The most energy we've really observed is in cosmic rays; more than LHC by circa 10^6, I think, but still not close to the Planck energy. Being a string theorist, I'm tempted to say something stringy would happen—but the truth is, we just don't know."

Luckily, that's not the end of the story. Remember how earlier we decided to ignore general relativity? Well, this is one of the rare situations in which bringing in general relativity makes a problem *easier* to solve.

There's a huge amount of potential energy in this scenario—the energy of all those electrons straining to be far apart from one another. That energy warps space and time just like mass does. The amount of energy in our electron Moon, it turns out, is about equal to the total combined mass and energy of the entire visible universe.

An entire universe worth of mass-energy—concentrated into the space of our (relatively small) Moon—would warp space-time so strongly that it would overpower even the repulsion of those 10^{52} electrons.

Dr. Keeler's diagnosis: "Yup, black hole." But this is no ordinary black hole; it's a black hole with a lot of electric charge.* And for that you need a different set of equations—rather than the standard Schwarzschild equations, you need the Reissner–Nordström ones.

The Reissner-Nordström equations compare the balance between the outward force of the electric charge and the inward pull of gravity. If the outward push from the charge is large enough, it's possible the event horizon surrounding the black hole can disappear completely. That would leave behind an infinitely dense object from which light *can* escape—what's called a naked singularity.

Once you have a naked singularity, physics starts breaking down in very big ways. Quantum mechanics and general relativity give absurd answers, and they're not even the *same* absurd answers. Some people have argued that the laws of physics simply don't allow for that kind of situation to arise. As Dr. Keeler put it, "Nobody likes a naked singularity."

In the case of an electron Moon, the energy from all those electrons pushing on one another would be so large that the gravitational pull would win, and our singularity would form a normal black hole. At least, "normal" in some sense; it would be a black hole as massive as the observable universe.†

---

* The proton Earth, which would also be part of this black hole, would reduce the charge, but since an Earth-mass of protons has much less charge than a Moon-mass of electrons, it doesn't affect the result much.

† A black hole with the mass of the observable universe would have a radius of 13.8 billion light-years, and the universe is 13.8 billion years old, which has led some people to say "the universe is a black hole!" This sounds like some kind of deep insight, but it really isn't true. The universe isn't a black hole. For one thing, everything in it is flying apart, which is something black holes famously don't do.

Would this black hole cause the universe to collapse? Hard to say. The answer depends on what the deal with dark energy is, and *nobody* knows what the deal with dark energy is.

But for now, at least, nearby galaxies would be safe. Since the gravitational influence of the black hole can only expand outward at the speed of light, much of the universe around us would remain blissfully unaware of our ridiculous electron experiment.

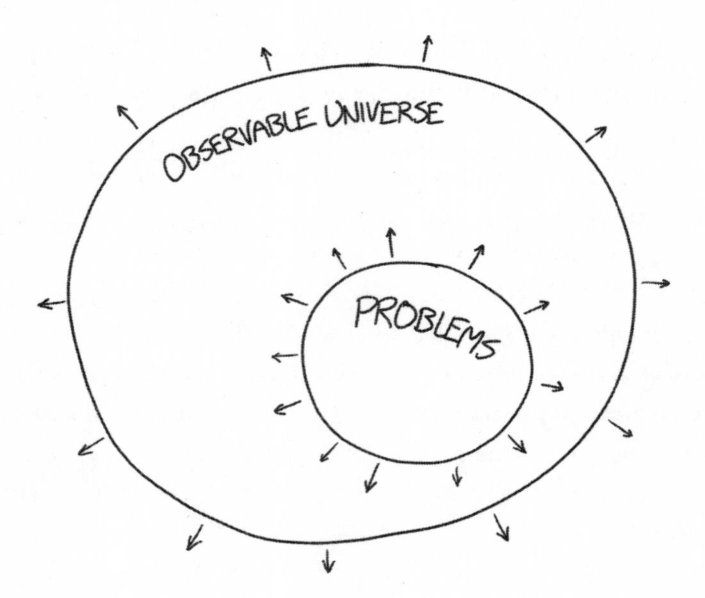

49. EYEBALL

If I pulled out my eyeball and aimed it so that it was looking into my other eyeball, what would I see (assuming the nerves and veins remain undamaged)?

—Lenka, Czech Republic

You would see an eyeball. The eyeball would be surrounded by a haze of double vision, where you'd see an overlapping face and a hand superimposed over the background of your room.

Pointing an eyeball at an eyeball doesn't create some kind of weird loop, like pointing a camera at its own video feed. Each eyeball just sees an eyeball. If you managed to line them up carefully, the two eyeballs would overlap, and your brain would try to combine the two similar images, the way it normally does when you look at a scene through two eyes.

Outside of the pupil and iris in the center of your vision, your two eyes would see totally different scenes. One eye would see an eyelid, a head, and one side of the room you're in. The other eye would see an eyeball, a hand, an optic nerve, and the other side of the room. Your brain wouldn't be able to combine these two overlapping images at all, so you'd have double vision everywhere outside of a small area at the center.

As I've mentioned, I'm not a medical professional, so take this advice with a grain of salt, but I don't think you should remove your own eyeball.

If you don't want to perform barehanded ophthalmological surgery,* you can get an idea of what you would see in this scenario using a mirror. If you put a regular mirror up to your face and stare ahead, each eyeball will be looking into itself, much like what would happen in your eye-removal scenario. To mimic it more closely, you could use a pair of mirrors at a right angle, so each eye is looking into the other, the way it would be if you held your eye in front of you.

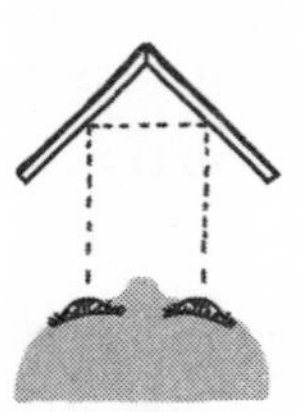

* For some reason.

If you try this, you'll notice that your eyes can't focus closer than a few inches, which is a limitation of the lens of your eye. This minimum focus distance increases with age, from 2 or 3 inches for children to 6 inches by age 30 or 40, and up to several feet by age 60 or 70. But regardless of your age, you'll need a magnifying glass or some very strong reading glasses to hold the mirrors close enough to see your eyes in detail. Extra light will also help, since the mirrors will block the light from the room.

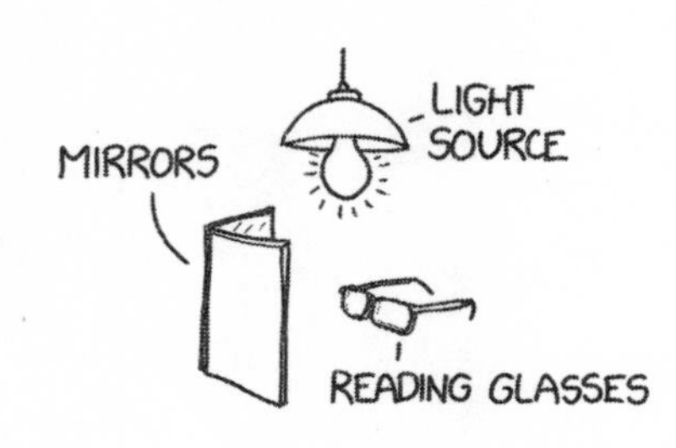

Since your eyes aren't symmetrical, the two images you see won't line up. With the right-angle mirrors, your right eye will see an eye with the plica semilunaris—the little fleshy membrane at the corner of your eye next to your nose*—on the left side of the image. Your left eye will see the reverse. Even if your irises are symmetrical and free of colored blotches, you'll still get double vision around the edges.

It does look kind of neat—I tried it while writing this—but definitely doesn't seem like an experience worth removing an eyeball for. The eyes may be the windows to the soul, but if you want to gaze into yours, I would stick to mirrors.

* Birds have a nictitating membrane, a transparent "third eyelid" that they can blink to protect and moisturize their eye. Many other animals have them, although humans and our evolutionary relatives have lost them. That bit in the corner of your eyelid is the vestigial remnant of your nictitating membrane.

If ALL of Japan's islands disappear, would it affect Earth's natural phenomena (plates, oceans, hurricanes, climate, and so on)?

—Miyu Uchida, Japan

The islands of Japan form a volcanic arc, with the Sea of Japan/East Sea on one side and the Pacific Ocean on the other.

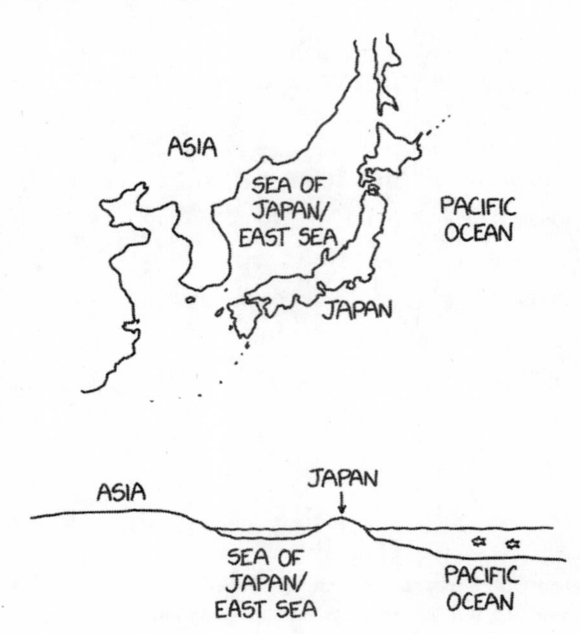

I'm not sure what sort of disappearance Miyu is planning, but let's assume the whole archipelago just kind of zips away somewhere for a while to run an errand.

Japan—the part of it that's above sea level—weighs 440 trillion tons. If just that part teleported away . . .

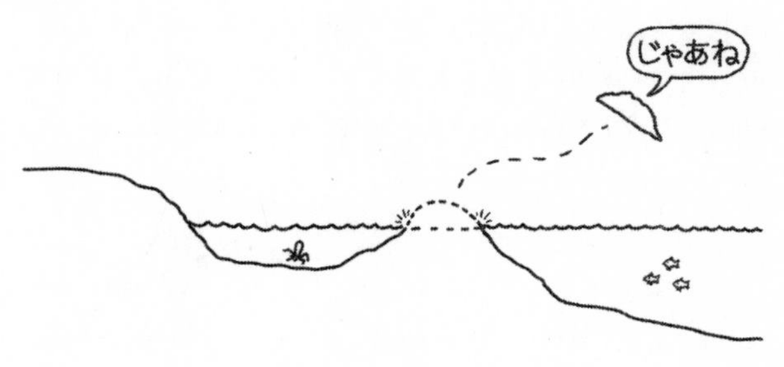

. . . it would shift the Earth's center of mass and axis of rotation toward Uruguay—the opposite side of the Earth—by about a foot and a half.

The change in the pull of gravity would cause the oceans to slosh around a little, with the sea settling on a new "sea level" that follows the contours of the new geoid. Without Japan's gravity, the ocean would shift slightly toward the other side of the Earth; sea level would probably fall by a foot or two around east Asia, and rise by the same amount around South America.*

* This effect also happens when big ice sheets on land melt. Their water causes the sea level to rise overall, but since their gravity no longer pulls the ocean *toward* them, the sea level can actually *fall* in the area around the sheet. On the other side of the world, it will rise by more than you'd expect. If or when Greenland melts, the flooding will be worst in Australia and New Zealand. For more on this, see *How To,* chapter 2, "How to Throw a Pool Party."

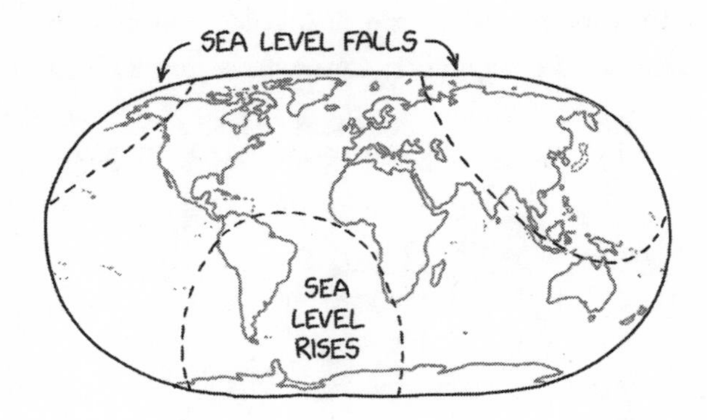

This foot and a half of sea-level rise would have pretty dramatic impacts on Uruguay, submerging lots of coastline. Of course, we don't need a hypothetical scenario for that, since that's how much the seas will rise over the next half century or so already thanks to human greenhouse-gas emissions.

So far, we've only considered removing the part of Japan that's *above* sea level. What about the rest of Japan? What if we remove the underwater part, too?

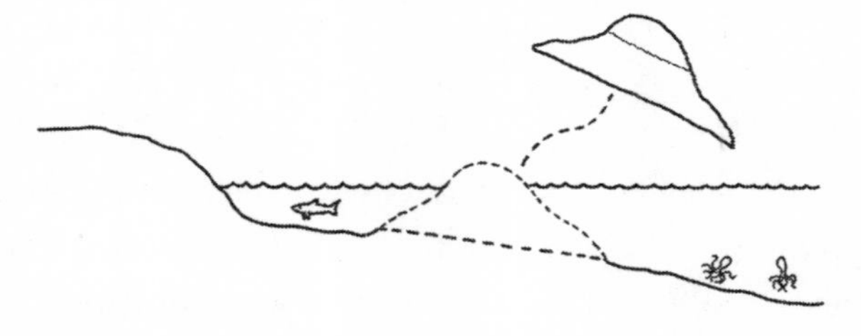

This part of Japan outweighs the above-water portion by more than 10 to 1.

DID YOU KNOW THAT ONLY 10 PERCENT
OF JAPAN IS VISIBLE? 90 PERCENT OF IT
IS HIDDEN BELOW THE OCEAN'S SURFACE!

If you removed the below-water portion of Japan, the shift in the Earth's axis would be much larger—10 or 20 feet—and so would the readjustment of sea level.

Removing Japan would also have a big effect on ocean currents. The sea to the west of Japan is linked to the surrounding oceans by just a few shallow straits, so the water in it is relatively isolated. It has its own circulation that keeps the layers of water well mixed; it resembles a miniature version of a larger ocean like the north Atlantic. Without the islands of Japan to cradle it, the sea would mix freely into the Pacific.

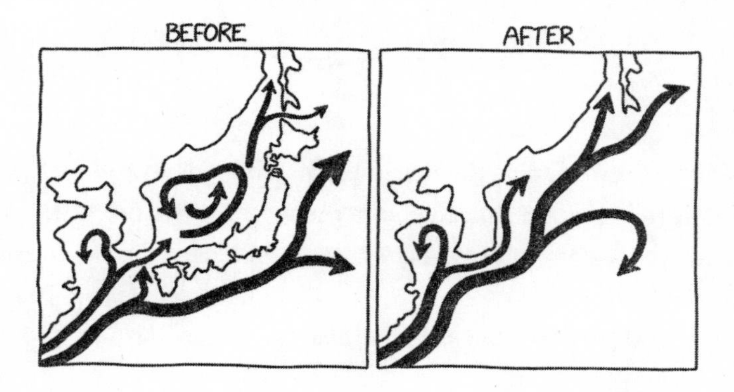

The effect on the climate would be hard to predict. Japan is warmed by the Kuroshio Current, which brings warm water up along the western edge of the Pacific, skirting the eastern side of the islands. With that barrier gone, the current would probably hug the coast of Asia, which would mean warmer water near Vladivostok and possibly a slightly increased risk of typhoons along the Korean peninsula and the Russian coast. However, they wouldn't need to worry about storm surge, since the sea level would have fallen and left the glass beaches of Vladivostok* high and dry.

At least, they wouldn't need to worry about storm surge in the *long* term. If Japan disappeared down to the sea floor, it would leave a giant cavity in the ocean. The ocean would rush in to fill the cavity, creating a bigger splash than any seen on Earth since the last giant space impact.† The wave would devastate the west coast of Asia,

* If you're not familiar with it, I recommend doing a quick image search for "glass beaches of Vladivostok"—you won't regret it!

† The last impact tsunami on that scale happened when a space rock hit the east coast of North America 35 million years ago. I went to school at Christopher Newport University in Virginia, which is built on the rim of the buried crater left by the impact.

and even when it crossed the Pacific, it would still be large enough to inundate the west coasts of the Americas and crash against the Andes and the Sierra Nevada.

When the water returned to the ocean basins, the seas would be lower than they had been, thanks to the Japan-shaped gap in the western Pacific. When Japan returns from its errand, if it wants to settle back into its old spot, it risks causing the same cataclysm all over again.

But then again, Miyu never did say where Japan was going.

Maybe the move was permanent.

51. FIRE FROM MOONLIGHT

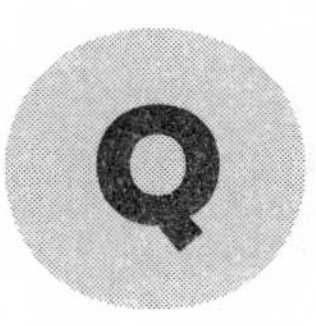

Can you use a magnifying glass and the moonlight to light a fire?

—Rogier

At first, this sounds like a pretty easy question.

A magnifying glass concentrates light on a small spot. As many mischievous kids can tell you, a magnifying glass as small as a square inch in size can collect enough light to start a fire. A little googling will tell you that the Sun is 400,000 times brighter than the Moon, so all we need is a 400,000-square-inch magnifying glass. Right?

Here's the real answer: **You can't start a fire with moonlight*** no matter *how* big your magnifying glass is. The reason is kind of subtle. It involves a lot of arguments that sound wrong but aren't, and generally takes you down a rabbit hole of optics.

* Pretty sure this is a Bruce Springsteen song.

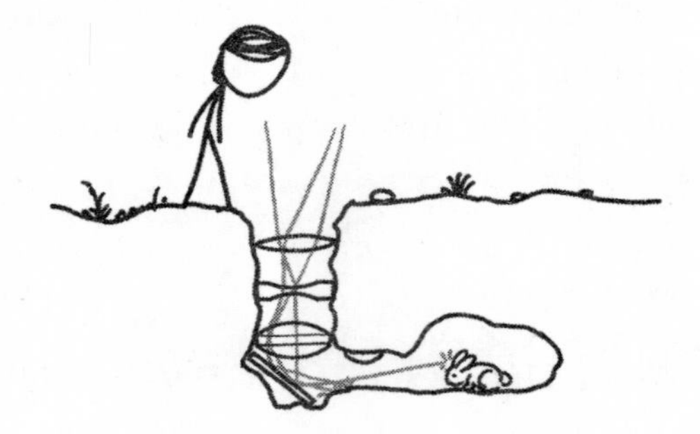

First, here's a general rule of thumb: **You can't use lenses and mirrors to make something hotter than the surface of the light source itself.** In other words, you can't use sunlight to make something hotter than the surface of the Sun.

There are lots of ways to show why this is true using optics, but a simpler—if perhaps less satisfying—argument comes from thermodynamics:

Lenses and mirrors work for free; they don't take any energy to operate.* If you could use lenses and mirrors to make heat flow from the Sun to a spot on the ground that's hotter than the Sun, you'd be making heat flow from a colder place to a hotter place without expending energy. The second law of thermodynamics says you can't do that. If you could, you could make a perpetual motion machine.

* More specifically, everything they do is fully reversible—which means they don't increase the entropy of the system.

The Sun is about 5,000°C, so our rule says you can't focus sunlight with lenses and mirrors to get something any hotter than 5,000°C. The Moon's sunlit surface is a little over 100°C, so you can't focus moonlight to make something hotter than about 100°C. That's too cold to set most things on fire.

"But wait," you might say. "The Moon's light isn't like the Sun's! The Sun is a blackbody—its light output is related to its high temperature. The Moon shines with reflected sunlight, which has a 'temperature' of thousands of degrees, so that argument doesn't work!"

It turns out it *does* work, for reasons we'll get to later. But first, hang on—is that rule even correct for the Sun? Sure, the thermodynamics argument seems simple enough, but to someone with a physics background who's used to thinking of energy flow, it may sound a little puzzling. Why *can't* you concentrate lots of sunlight on to a point to make it hot? Lenses can concentrate light down to a tiny point, right? Why can't you just concentrate more and more of the Sun's energy down on to the same point? With over 10^{26} watts available, you should be able to get a point as hot as you want!

Except lenses *don't* concentrate light down on to a point—not unless the light source is also a point. They concentrate light down on to an *area*, creating a tiny image of the Sun.* This difference turns out to be crucial. To see why, let's look at an example:

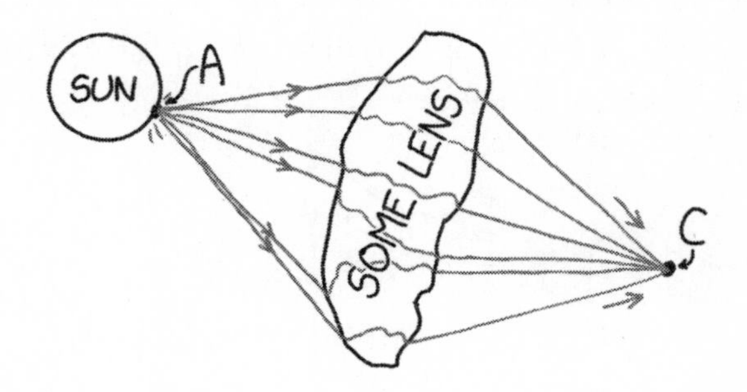

* Or a big one—some home telescopes, such as the wood-framed Sunspotter, use lenses to project a detailed image of the Sun onto a sheet of paper, like a high-resolution version of a pinhole camera. They're a little expensive, but they're a great tool for safely viewing sunspots or solar eclipses.

This lens directs all the light from point A to point C. So far, so good. But if we're saying this lens concentrates all the light from the Sun down to a point, that means it must direct all the light from point B to point C, too:

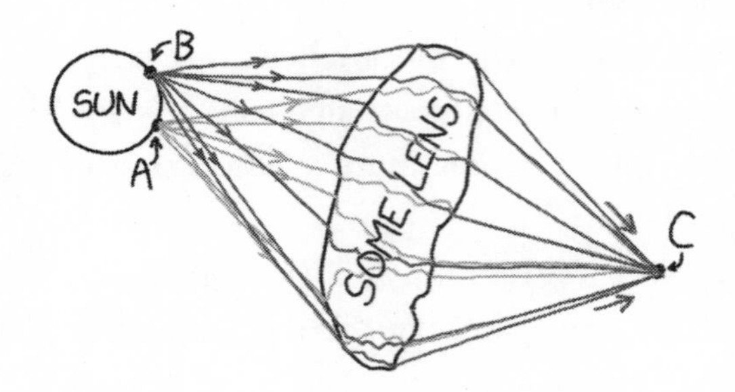

Now we have a problem: What happens if you shine rays of light from point C back toward the lens? Optical systems are reversible, so the light should be able to go back to where it came from—but how does the lens know whether the light came from B or A?

In general, it turns out there's no way to "overlay" light beams on each other, because it would violate the reversibility of the system. This rule keeps you from sending more beams of light toward a target from the same direction as existing ones, which puts a limit on how much light you can direct from a source to a target.

Maybe you can't overlay light rays, but what about sort of, you know, smooshing them closer together, so you can fit more of them side by side? Then you could gather lots of smooshed beams and aim them at a target from slightly different angles.

Nope, you can't do this, either.*

* We already know this, of course, since earlier we said that it would let you violate the second law of thermodynamics.

It turns out that any passive optical system follows a law called "conservation of étendue." This law says that if you have light coming into a system from a bunch of different angles *and* over a large "input" area, then the input area times the input angle* equals the output area times the output angle. If your light is concentrated to a smaller output area, then it must be "spread out" over a larger output angle.

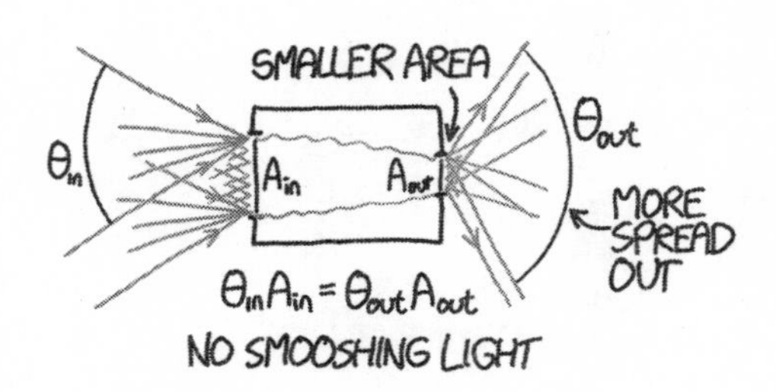

In other words, you can't smoosh light beams together without also making them less parallel, which means you can't aim them at a faraway spot.

There's another way to think about this property of lenses: They only make light sources take up more of the sky; they can't make the light from any single spot brighter. You can see this by holding a lens up to a wall and looking at it. No matter what kind of lens you use, you'll find it doesn't make any part of the wall look brighter; it just changes which part of it you see in that direction. It can be shown† that making a source of light brighter would violate the rules of étendue, so it's impossible for a lens system to do that. All it can do is make every line of sight end on the surface of a light source, which is equivalent to making the light source surround the target.

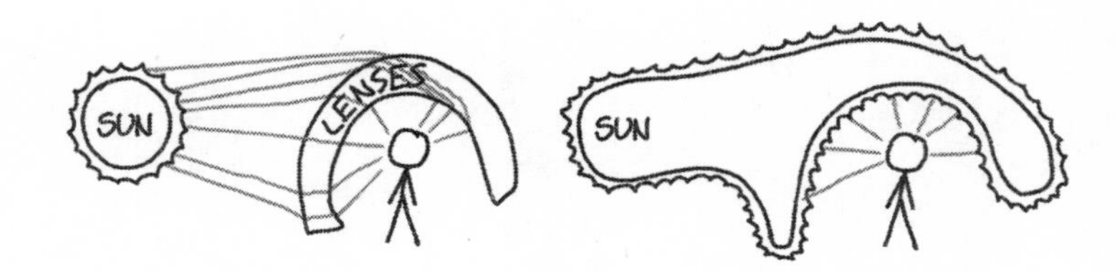

* Or *solid angle,* in 3D systems.

† This is physics-speak for "this probably isn't too hard, but I don't want to do it."

If you're "surrounded" by the Sun's surface material, then you're effectively floating within the Sun, and will quickly reach the temperature of your surroundings.*

If you're surrounded by the bright surface of the Moon, how hot will you get? Well, rocks on the Moon's surface are nearly surrounded by the surface of the Moon, and they reach the temperature of the surface of the Moon (since they *are* the surface of the Moon). So a lens system focusing moonlight can't really make something hotter than a rock sitting in a small depression on the Moon's surface.

Which gives us one last way to prove that you can't start a fire with moonlight: The Apollo astronauts survived.

SO WHEN THE LAST APOLLO ASTRONAUT DIES...
PHYSICS WILL COLLAPSE AND THE MOON WILL START FIRES.

* See chapters 61, 62, and 63, along with Short Answers #5, for more on the exciting experiences you can have by visiting the Sun.

52. READ ALL THE LAWS

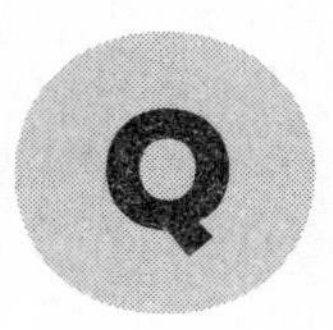

If a person wanted to read all of the governing documents that apply to them—from the federal and state constitutions, treaties, agency-issued regulations, federal and state laws, local ordinances, etc.—how many pages would they have to read?

—Keith Yearman

There are a lot of laws. To find out what's in them, you have to read them. Otherwise, you could commit a crime and not know it. For all you know, any of your seemingly ordinary hobbies or activities might violate some obscure law.

I'M JUST A REGULAR PERSON HANGING OUT AROUND MY HOME. MY HOBBIES INCLUDE GARDENING, JOGGING, CATCHING AND EATING MIGRATORY BIRDS, DRILLING FOR OIL, BUYING AND SELLING LASERS, LAUNCHING INCREASINGLY LARGE MODEL ROCKETS, AND MAKING RANDOM DEFAMATORY CLAIMS IN THE TOWN SQUARE.
I HOPE I DON'T GET IN ANY LEGAL TROUBLE!

I live in a town in Massachusetts, so I'm under the jurisdiction of the following governing documents:

- The US constitution (26 pages)
- Federal laws (82,000 pages*)
- The Massachusetts constitution (122 pages)
- Massachusetts state laws (63,000 pages)
- My town's laws (450 pages)

That's a total of about 145,000 pages. If you read at 300 words per minute for 16 hours a day, they would take about six months to read.

But those are just the *statutes*, or laws passed by the legislature. In addition to those, there are also regulations, rules issued by authorized governmental organizations. These are often published alongside laws and include, among other things:

- Federal regulations (295,000 pages)
- Massachusetts regulations (31,000 pages)
- My town's municipal zoning regulations (500 pages)

Including these regulations† more than triples how much we have to read, bringing the total reading time up to almost two years.

Article VI of the US Constitution adds in another source of law—treaties.

> *This Constitution, and the Laws of the United States which shall be made in Pursuance thereof;* **and all Treaties made**, *or which shall be made, under the Authority of the United States, shall be the supreme Law of the Land . . .*

—Article VI

* In some cases I'm using the actual page counts and in others I'm using word counts and assuming 350 words per page, which is typical for printed legal documents.

† There are other rules, like electrical codes, that are "incorporated by reference." A law might say something like "if you sell a crazy straw, it has to comply with the Crazy Straw Standard 385-1.2 published by the National Crazy Straw Manufacturers' Organization." These are things you might have to read to interpret laws, but they don't really count as a primary source of law themselves, so we'll skip them.

The US State Department publishes an annual list of all the United States's active treaties and agreements. The 2020 list is 570 pages long. That's not how long the treaties are, that's just how long the *list* of treaties is. At about 14 treaties per page, that gives a total of 7,700 treaties. In January 2005—picking a random time period to sample—the average length of a treaty was 33 pages. If that average applies to the whole set, that adds up to a quarter of a million pages, bringing our total to about 700,000 pages, which will take 2½ years to read.

Last, but not least, there's *case law*. When the Supreme Court "strikes down" a law, the law isn't actually deleted. The court just says that it can no longer be enforced, and sometimes orders people or law enforcement to act in a different way.* But the court doesn't actually make changes to the text of the law itself, so someone reading the original law might not know it's been struck down or altered by a court. If you want to know about these "updates," you have to read the court decisions, and it turns out there are a lot of them.

Massachusetts state case law totals about half a million pages, which will add another two years to your total reading time. Federal case law dwarfs all those sources of law, contributing a whopping 12.3 million pages. Reading all of it—even from the other federal districts, in case one of them issued a nationwide injunction that binds you—would take 41 years, for a grand total of 45 years.†

* Sometimes, this simply takes the form of "striking down" a law, but sometimes it *expands* the law.

† Depending on your view of nationwide injunctions, you might be able to get away with reading only the Supreme Court's cases and the ones in your own district, which would drop it down to a more manageable—but still probably impossible—7 years of reading.

DO I HAVE TO READ ALL THESE LAWS?

Most of the laws don't apply to you. For example, 42 U.S. Code § 2141(b) sets limits on the Department of Energy's ability to distribute nuclear materials. If you're not the Department of Energy, you don't need to worry about that.*

But there's not really any way to know which laws apply to you without reading them. If you don't know what the law is, there are plenty of activities that could get you in trouble. For example, California's Food and Agricultural Code § 27637 bars anyone from making false or misleading statements about eggs. Luckily, I don't live in California, so I'm free to share my egg theories.

* To those of you who *are* at the Department of Energy, hi! I'm a huge fan of your work, and of energy in general.

OKAY, BUT REALLY, HOW ARE YOU SUPPOSED TO KNOW WHAT'S ILLEGAL?

To get some answers, I reached out to the Harvard Law Library and asked research librarian A. J. Blechner how I—a humble citizen who just wants to partake in normal hobbies like rocketry or defamation—am supposed to know what's legal and what isn't.

"Your public law library could help you to find things," Blechner told me. Additionally, trial courts often have their own libraries, which are open to the public. "These were created to help judges and attorneys, but as a member of the public, you can just walk in and get some assistance. It's a great, not-that-well-known resource."

Law libraries are a great resource for learning about the law, but if you're worried that you might be in legal trouble, Blechner also has some more practical advice. "If you have a legal question that you're not sure how to answer," they told me, "talking to a lawyer is probably a good course of action."

DO WE REALLY NEED ALL THESE LAWS?

Laws give people power. If a law is complicated, it empowers people who can afford lawyers to interpret it. "Laws that are complicated, arbitrary, and unintuitive empower the state," says Jonathan Zittrain, professor of international law and the Harvard Law Library's director, "since prosecutorial discretion means they can pick whom to enforce against and be selective in discriminatory ways."

But making laws simpler and vaguer doesn't necessarily move that power from the state to the people. You could get rid of a lot of laws and replace them with "everyone just needs to behave properly." But that leaves it up to law enforcement to decide the meaning of "properly."

In a sense, the law is infinite in length, because it includes not just the words themselves but society's understanding of what those words mean. California says I can't share false or misleading information about eggs. If I say that you can hatch a real live Pikachu by incubating a poké ball, that's a false statement, but is it a statement about eggs? *Are* poké balls a type of egg?

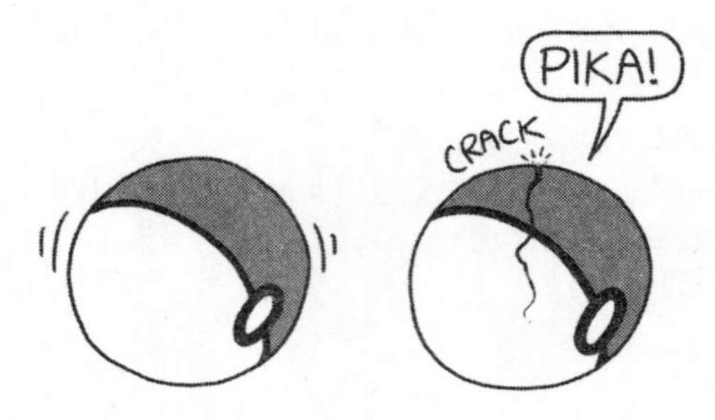

I don't think poké balls are eggs. But maybe most people think they are, and I just don't know about it because I'm not that into Pokémon. It might determine what is or isn't against the law, but the question of whether a poké ball counts as an egg is not clarified in the text of the law. At least, not as of this writing.

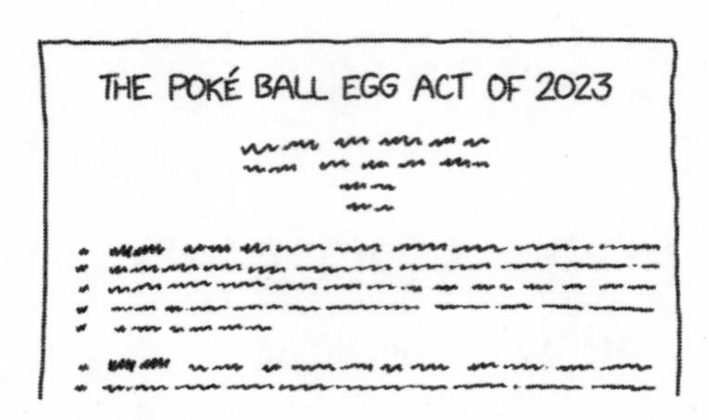

YOUR OWN LAW

What if you've finished reading all the laws, but you're having so much fun that you don't want to stop?

In some cases, Zittrain says you can create additional law just by requesting a clarification from the government. "In tax law, you can write a letter to the IRS to ask them if something you want to do will break the law. Their reply is a little more law just for you!"

So if you want a personalized piece of law, you can contact the IRS to request a *private letter ruling* and get a binding ruling in response. The IRS generally charges a fee for this service—which can be substantial, depending on how much work is involved—but at the end you'll have your very own official piece of law answering whatever question you've been wondering about.

HELLO, IRS COMMISSIONER? FOR TAX PURPOSES, ARE POKÉ BALLS CONSIDERED POULTRY PRODUCTS?
...I GOT THE NUMBER FROM YOUR WEBSITE. WHY?
...WELL, MAYBE YOU NEED TO SPEND MORE TIME PLAYING VIDEO GAMES, DID YOU EVER—
CLICK
WOW, RUDE.

weird & worrying #3

Q If I were to jump into a container of liquid nitrogen (or dispose of a body in that way), how deep would it have to be for me/them to shatter into frozen pieces at the bottom?

—Stella Wohnig

Q What would happen to you if a colony of ants suddenly appeared in your bloodstream all at once?

—Matt, on behalf of his son Declan, age 8

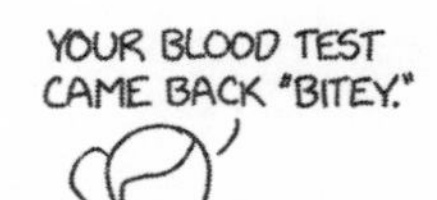

Q If Harry Potter forgets where the invisible entrance to Platform 9¾ is, how long would he have to crash into walls randomly before discovering it?

—Max Plankar

53. SALIVA POOL

How long would it take for a single person to fill up an entire swimming pool with their own saliva?

—Mary Griffin, ninth grade

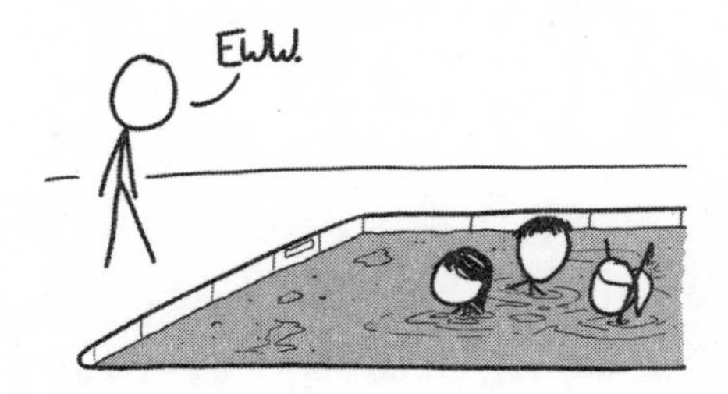

The average kid produces about half a liter of saliva per day, according to the paper "Estimation of the Total Saliva Volume Produced Per Day in Five-Year-Old Children," which I like to imagine was mailed to the *Archives of Oral Biology* in a slightly sticky, dripping envelope.

A 5-year-old probably produces proportionally less saliva than a larger adult. On the other hand, I'm not comfortable betting that *anyone* produces more drool than a little kid, so let's be conservative and use the paper's figure.

If you're collecting your saliva,* you can't use it to eat.† You could get around this by chewing gum or something, to get your body to produce extra saliva, or just by drinking liquid food or getting an IV.

At the rate of 500 mL per day from the paper, it would take you about a year to fill a typical bathtub.

Side effects of filling a bathtub with saliva include: dry mouth.

A bathtub full of saliva is pretty gross, but that's not what you asked about. For some reason—I don't really want to know why—you asked about filling a pool.

Let's imagine an Olympic-size swimming pool, which is 25 meters by 50 meters. Depths vary, but we'll suppose this one is uniformly 4 feet deep,‡ so you can probably stand up in it.

At 500 mL per day, it would take you 8,345 years to fill this pool. That's a long time for the rest of us to wait, so let's imagine you went back in time to get started on this project early.

---

* This question is gross, by the way.

† I hope.

‡ The Fédération Internationale de Natation's website says that a pool with starting blocks does need a slightly deeper bit near each end, but it can be shallower in the middle. There doesn't seem to be anything in the rules about a maximum depth, so I suppose you can make a pool that continues through to the other side of the Earth, but then you run into trouble when you try to follow the instructions in section FR 2.14 about painting lane markings on the bottom.

Eight millennia ago, the ice sheets that covered much of the northern parts of the world had mostly receded, and humans had just begun to develop agriculture. Let's imagine you started your project then.

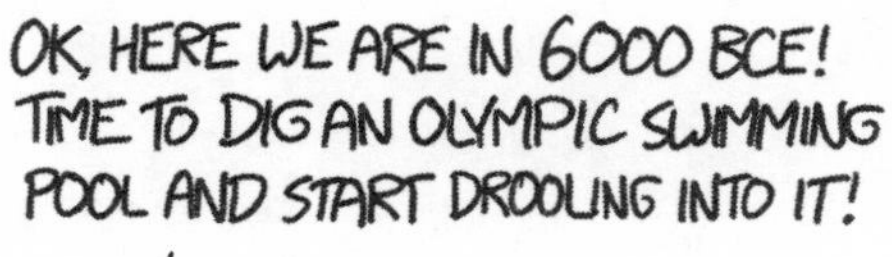

By 4000 BCE, when the civilizations of the Fertile Crescent had begun to develop in modern-day Iraq, the saliva would be a foot deep, covering your feet and ankles.

By 3200 BCE, when writing was first developed, the saliva would creep past your knees.

Around the mid-2000s BCE, the Great Pyramid was constructed and early Mesoamerican cultures were emerging. At this point, the saliva would be getting close to your fingertips if you didn't lift your arms up.

Around 1600 BCE, the eruption of a huge volcano in the Greek island now known as Santorini caused a massive tsunami that devastated the Minoan civilization, possibly causing its final collapse. As this happened, the saliva would probably be approaching waist-deep.

The saliva would continue to rise throughout the next three millennia of history, and by the time of Europe's industrial revolution it would be chest-deep, easily enough saliva to swim in. The last 200 years would add the final 3 centimeters, and the pool would finally be filled.

It would take a long time, sure. But it would all be worth it, because at the end of it all, you'd have an Olympic-size swimming pool full of saliva. And isn't that, deep down, all any of us really want?*

* No. It is not.

54. SNOWBALL

What if I tried to roll a snowball from the top of Mount Everest? How big would the snowball be by the time it reached the bottom and how long would it take?

—Michaeline Yates

When snowballs roll through wet, sticky snow, they grow. For dry snow like what you'd find on Mount Everest, a rolled snowball wouldn't get bigger; it would just tumble down the mountain like any other object.

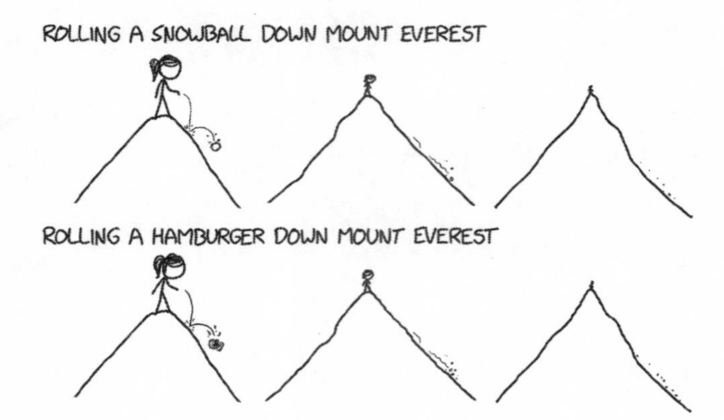

But even if Mount Everest were covered in the kind of wet snow that made for good snowballs, a snowball wouldn't get that big.

A rolling snowball picks up snow and gets bigger, and a bigger snowball picks up more snow. This may sound like a recipe for some kind of exponential growth, but an idealized snowball's growth actually slows down over time. It keeps getting bigger and wider, but each new meter it rolls adds less to the diameter. The growth slows because the width of the snowball's track—and thus the amount of snow it picks up—is proportional to its radius, but the surface area the new snow has to cover is proportional to radius squared, which means that each new clump of snow has to be spread out over more area. People use the word "snowballed" to mean "grew faster and faster," but in a sense the truth is the reverse.

Mount Everest is very tall,[citation needed] so even with a slowing growth rate, there's still a lot of room for a snowball to pick up snow. The mountain's three main faces descend about 5 kilometers before they level off into glacial valleys. In theory, an idealized snowball rolling down a 5-kilometer slope would pass through enough snow to grow to 10 or 20 meters wide by the time it reached the bottom.

In practice, it wouldn't make it more than a few hundred meters, even in perfect wet snow. There's a limit to how big snowballs can get before they collapse under their own weight. Gravity pulls the edges of a snowball down, so the insides are under tension. If a snowball gets too big, it collapses.

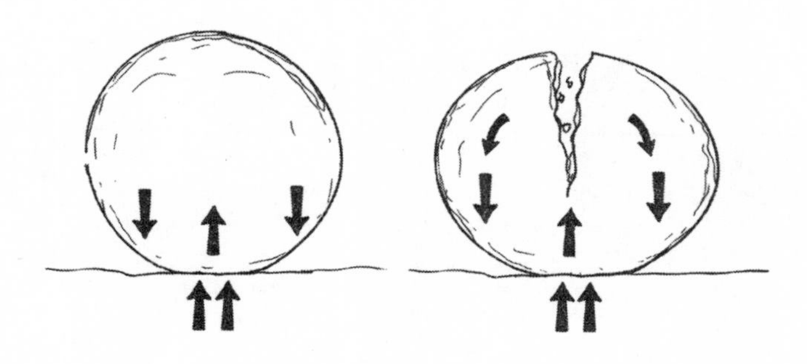

Snow has a tensile strength, which means it resists being pulled apart. Its tensile strength isn't *that* high—which is why you don't see a lot of ropes made of snow—but it's not zero. A typical tensile strength for well-packed snow might be a few kilopascals, which is stronger than wet sand, weaker than most kinds of cheese, and about 1/10,000th that of most metals.

There's a number in engineering that measures how long a dangling piece of a material can get before snapping under its own weight. It's called the "free-hanging length," and it's a ratio between a material's tensile strength, density, and gravity.

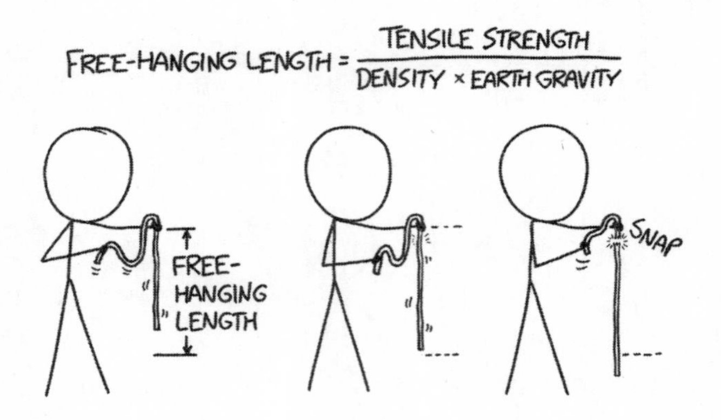

The free-hanging length of a material provides a pretty decent approximation—to within an order of magnitude, at least—of how big a ball of that material could get. Its value for snow ranges from less than a meter for fluffy snow to a meter or two for heavy, packed snow.

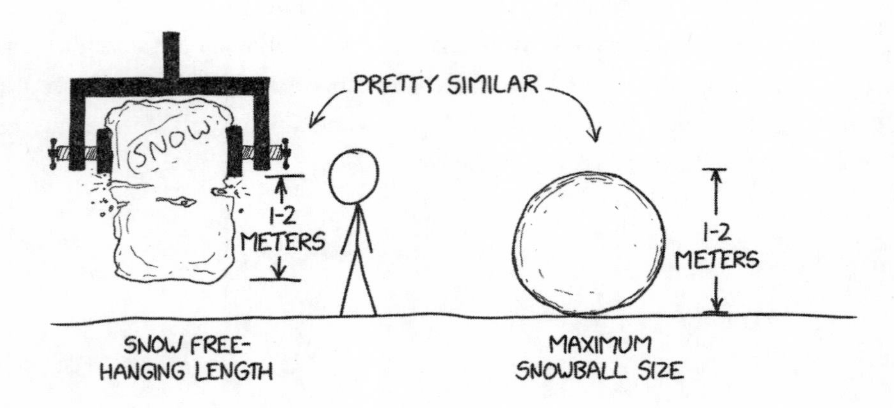

This formula lets us compare different materials. It tells us that the largest snowball would be bigger than the largest ball of sand—which is even weaker than snow and much more dense—but smaller than the largest ball of hard cheese and nowhere near as large as the largest ball of iron.

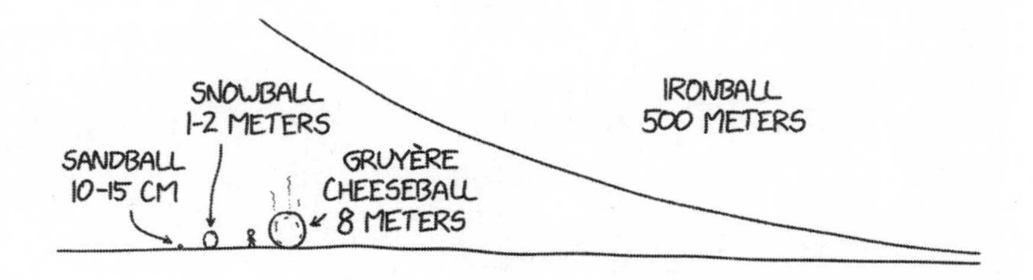

If you look up videos of people rolling large snowballs down hills, you'll see that they usually break apart when they reach a size of a few meters, just as the formula suggests.

But slopes that can support self-growing snowballs are rare, and they're rare *because* they can support self-growing snowballs. If a snowball grows while it's rolling down a hill, it will break apart. A snowball that breaks apart becomes a bunch of little snowballs, which will start to grow, too, just like the original.

Congratulations, you've invented an avalanche.

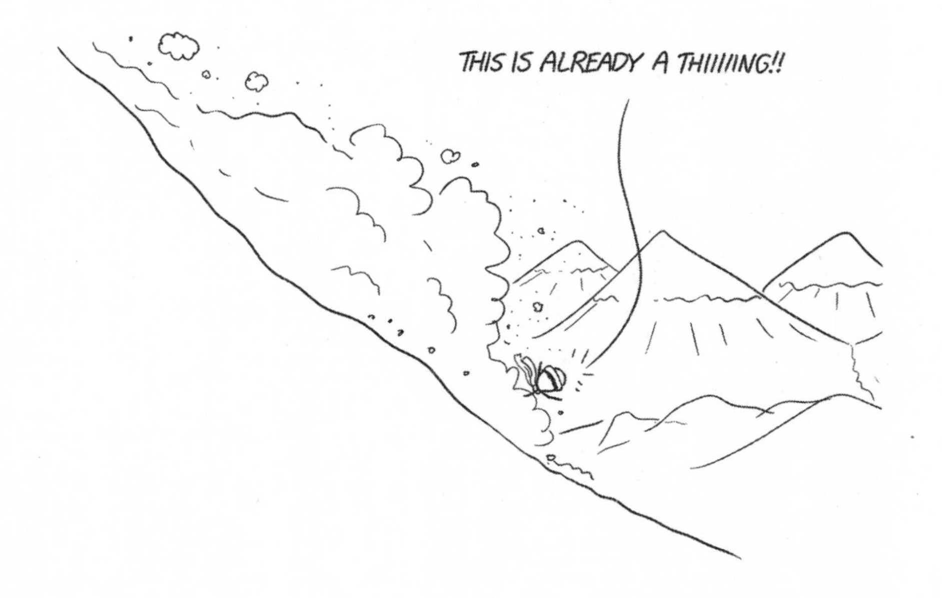

55. NIAGARA STRAW

What would happen if one tried to funnel Niagara Falls through a straw?

—David Gwizdala

One would get in trouble with the International Niagara Committee, the International Niagara Board of Control, the International Joint Commission, the International Niagara Board Working Committee, and probably the Great Lakes–St. Lawrence River Adaptive Management Committee.* Also, the Earth would be destroyed.

Well, that's not quite right. At the risk of stating the obvious, the real answer is, "Niagara Falls wouldn't fit through a straw."

---

* Which is, if I'm understanding these organizational charts right, itself a supergroup made up of three committees for individual bodies of water.

There are limits to how fast you can push fluids through things. If you pump a fluid through a narrow opening, it speeds up. If the fluid is a gas,* it becomes "choked" when the speed of the gas flowing through the opening reaches the speed of sound. At that point, the gas flowing through the hole can't move any faster—although you can still get more mass to flow through per second by increasing the pressure, which compresses the gas further.

For water, a different effect causes it to choke. When a fluid flows through an opening fast enough, the pressure within the fluid drops due to Bernoulli's principle. Water always "wants" to boil, but is held together by air pressure. When the pressure abruptly drops, bubbles of steam form in the water. This is called "cavitation."

When the water is forced through an opening at high speed, cavitation bubbles cause it to become less dense overall. Increasing the pressure—to try to push the water through harder—only makes it boil faster.† This keeps the total amount of water making it through the opening from rising, even if the water-steam mix moves at a higher speed.

Another limit on the water-flow rate comes from the speed of sound. You can't use pressure to accelerate water through an opening faster than the speed of sound (in water).‡ However, water very rarely reaches this point, because "the speed of sound (in water)" is very fast. Water is heavy, and if you try to make it go that fast, it tends to start ignoring the turns in your pipes.

So how fast does Niagara Falls need to go to fit through a straw, and is it faster than the speed of sound? This is easy to figure out; all we need to know is the flow rate over the falls and how much area it

* In physics, gases are considered a type of fluid.

† Valve designers try to avoid creating these steam bubbles, because after the bubbles form, they quickly collapse as the pressure rises back up on the other side of the valve, and the force from that collapse can gradually eat away at plumbing.

‡ It's sort of like a traffic jam—forcing more cars into the back of a traffic jam won't make the ones in the front come out faster. The analogy between traffic jams and choked flows isn't perfect, but I still like it because it's fun to imagine someone trying to solve traffic jams by using a bulldozer to push more cars into them.

needs to fit through, then we can divide the former number by the latter to give us the speed.

The flow rate over Niagara Falls is at least 100,000 cubic feet per second, which is actually mandated by law. The Niagara River supplies an average of about 292,000 cubic feet per second to the falls, but much of it is diverted into tunnels to generate electric power. However, since people get mad if you turn off the world's most famous waterfall, the generation facilities are required to leave at least 100,000 of those cubic feet per second flowing over the falls for everyone to look at (50,000 at night or during the off-season). There's periodic discussion of turning off the falls again for maintenance, and probably to see what cool stuff they can find while they're at it.

Important note: If you divert the water into a straw, you'll be in violation of the 1950 treaty that establishes the "100,000 cubic feet per second" limit.* This is monitored by the International Niagara Board of Control, which consists of one American and one Canadian.† They'll probably be upset with you, as will the other boards I mentioned earlier, so proceed at your own risk.

---

* As those of you who were inspired by chapter 52 to read all of the United States's laws and treaties already know, of course.

† As of 2021, the waterfall guardians are Aaron Thompson of Canada and Stephen Durrett of the United States. I'm guessing their enforcement protocol is just some variation on "filing a report," but I like to imagine that they're empowered to physically return the stolen water to the falls by any means necessary.

A typical straw is about 7mm in diameter. To find out how fast the water flows, we just divide the flow rate by that area. If the result is greater than the speed of sound, our flow will probably be choked, which will lead to problems.

$$\frac{100,000\,\frac{\text{cubic feet}}{\text{second}}}{\pi\left(\frac{7\text{ mm}}{2}\right)^2} = 73,600,000\,\frac{\text{meters}}{\text{second}} = 0.25c$$

Apparently, our water will be going one-quarter of the speed of *light*.

WATER SPEED, IN QUARTERS OF C	PROBLEMS?
0	MAYBE
1	YES
2	YES
3	YES
4	VERY YES
5	PLEASE STOP

On the plus side, we don't need to worry about cavitation since these water molecules would be going fast enough to cause all kinds of exciting *nuclear* reactions when they hit the walls of the straw. At those high energies, everything is a plasma anyway, so the concepts of boiling and cavitation don't even apply.

But it gets worse! The recoil from the relativistic water jet would be pretty strong. It wouldn't be enough to push the North American plate south, but it would be enough to destroy whatever device you were using to create the jet.

No machine could actually accelerate that much water to relativistic speeds. Particle accelerators can get things going that fast, but they're typically fed from a small bottle of gas. You can't just plug Niagara Falls into the accelerator input. Or, at least, if you did, the scientists would get awfully mad.

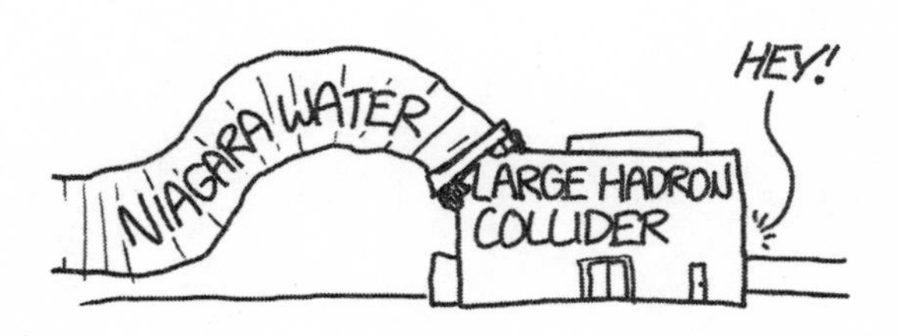

Which is for the best, since the power of the particle jet created by this scenario would be greater than the power of all the sunlight that falls on Earth. Your "waterfall" would have a power output equivalent to that of a small star, and its heat and light would quickly raise the temperature of the planet, boil away the oceans, and render the whole place uninhabitable.

And yet I bet someone would *still* try to go over it in a barrel.

56. WALKING BACKWARD IN TIME

What if you decided to walk from Austin, Texas, to New York City, but every step you take takes you back 30 days?

—Jojo Yawson

In the first *What If?* book, we imagined what you'd see if you stood in New York and jumped farther and farther backward in time. This question envisions a different kind of time-travel trip to New York.

As you lifted your foot to take your first step and time started running backward, the Sun would become a bright arch across the sky from horizon to horizon. The cars and pedestrians around you would vanish as human activity blurred into invisibility around you.

The Sun would become a strobe light in the sky. If you walked at a normal pace, 50 days would flash by each second, which means the world would cycle between light and dark at a frequency of 50 Hz. That frequency is right on the edge of the eye's "flicker fusion threshold," where flashes of light are too fast for your eyes to distinguish and appear to merge together into a steady glow, so the light would appear generally steady, if a little unnatural. The changing weather would add an extra layer of irregular flickering, as skies oscillated between periods of overcast and clear. Your eyes would get used to it after a while, I hope.

The Sun would appear as a band across the sky, like a fluorescent tube. It would move slowly up and down, once every 7 or 8 seconds, with the summer and winter cycles. Around you, trees would slowly retract toward the ground as you walked. With each annual cycle, branches of fruit trees would snap downward under the sudden weight of ripe fruit that jumped up from the ground, then gradually rise back up as the fruit unripened and retracted back into the branches.

Let's assume you start at the Texas state capitol in the middle of the city. From Austin, New York City is to the northeast, so you'd probably want to head toward the capitol complex's north exit. By the time you reached West 15th Street, at the edge of the green, it would be the year 2000.

Across the street to your right, the Robert E. Johnson Legislative Office Building would abruptly disassemble itself. As you crossed the street and walked down Congress Avenue, every 5 or 10 seconds a skyscraper would descend out of view like a prairie dog ducking into a burrow.

After ten minutes of walking, you'd reach the University of Texas at Austin campus somewhere in the mid-1940s. As you walked past the buildings, they would come apart and retract into the ground. By the time you were halfway across the campus, the university—established in 1883—would be gone.

As the university vanished, the railroads would disappear outside the city, and with them the millions of acres of plowed cropland that they supported. Within the span of a minute or two, sprawling farms would be replaced by open pastures. But these wouldn't be the native grasslands of modern pastures, which are mainly Bermuda grass and Bahia grass. This would be an entirely different ecosystem, a mix of grasses dotted with trees: America's lost prairies.

The violent removal of Indigenous people by Europeans would play out in reverse in an invisible blur around you. After half an hour of walking, the Europeans would be gone, and you would be among the Lipan Apache.

As you walk, pulses of fire would sweep the land, many of them set by people to encourage the maintenance of grasslands that supported herds of bison. The farms and towns of the Caddo nation would lie ahead to the northeast, but they won't be there by the time you reach them.

By the time you were 20 miles from Austin, you would be 4,000 years in the past. Corn and squash farms would become more rare as the development of agriculture unwound around you.

After you'd walked for 12 hours, there would be an ominous development. On the other side of the continent, in northern Quebec, a pancake-shaped sheet of ice would begin to grow and spread outward across the land. On the coast of Texas to your south, the sea—which would have gradually fallen by a few meters over the course of your walk—would abruptly withdraw from the coast, revealing hundreds of miles of grassy plains and forests.

As you reached the current site of Thorndale, Texas, after a full day of walking, large animals would proliferate around you. If you stopped walking for a moment, you might catch sight of a camel, a mastodon, a dire wolf, or a saber-toothed cat. A little past Thorndale, humans would disappear from the landscape entirely. We're not certain why all these big, cool animals vanished right around the time that humans arrived on the scene, but a lot of people suspect it might be more than a coincidence.

To the north, the expanding ice sheet would swallow much of the continent, but it wouldn't reach quite as far south as you, so you'd only feel the indirect effects as the climate changed around you.

After a week of walking, you would find yourself in Arkansas. Following its sudden incursion into the continent earlier in your walk, the ice would have slowly withdrawn back into Canada in fits and starts, and the sea would rise to cover the now-barren coastal lands. Around this same time, a supervolcano erupted in Sumatra, Indonesia, creating what is now Lake Toba. Some scientists have speculated that the eruption created a decade-long global winter and caused the human population to plunge, but the hypothesis is disputed. If you could pause for a minute and take some notes on what you see, a lot of researchers would really appreciate it.

After ten days of walking, you would reach the Mississippi River, slightly earlier than you might expect. The river is old—it's been here in one form or another for millions of years—but it moves around quite a lot, and you'd likely find it a bit to the west of its modern location. As you approached, you'd see it thrashing back and forth across a floodplain, with loops flinging themselves back and forth almost at walking speed, surrounded by a flickering blur from the periodic floods that submerge the plains all around you. Hopefully whatever process is keeping your low-speed lungs full of air also keeps you from drowning as you try to cross a river that's flowing at, from your point of view, 1 or 2 percent of the speed of light.

Assuming you managed to flail your way across the river, you'd find a much more Arctic landscape on the far side. Surprise—it's another ice age glaciation!

This one is the Illinoian glaciation, one of North America's most extreme glacial episodes. Your path would be a little too far south for the glaciers themselves to reach you, but before they expanded, the glacial floods would happen around you in reverse. Torrents of meltwater would periodically emerge from the ocean and rush north past you on their way to flow up onto the walls of ice and freeze in place.

During the week or so that it would take for you to cross the boreal spruce and jack-pine forests of Tennessee and Kentucky, the temperature would steadily rise. Around the 3-week mark, as you reached the Ohio River and the Appalachians, the climate would be downright warm. You would be at the peak of an interglacial period, 240,000 years ago, when temperatures were nearly as warm as they are today.*

As you crossed the Appalachians, the ice sheets would make one final lunge toward you, as part of what's called the MIS-8 glacial period, between 250,000 and 300,000 years ago. Your route would probably be far enough south to avoid them, but if you happened to take a more northerly path, you might encounter a pulsating wall expanding and retreating with the seasons. If you got too close, lobes of the ice sheet might occasionally surge forward with the speed of a freight train and a lot more momentum. Don't get too close.

---

* I'm writing this in the early twenty-first century.

As you approached New York City through the hills of northern New Jersey, you'd initially see a grassy plain, with rivers running out across it to the southeast. But as you drew closer, the distant sea would come into view. It would look like a long, slow tide advancing in fits and starts across the ground, sometimes as fast as walking speed. By the time you reach New York City, about 300,000 years in the past, the beach would be there to greet you, fairly close to its modern shoreline.

While the ocean might be in about the same place, the landscape of New York wouldn't be particularly recognizable. Familiar modern landmarks will have been scoured away by glaciers and re-formed by rivers in the intervening 300,000 years.

In the first *What If?*, the reader stands in New York City and skips backward through time, jumping from 100,000 years ago to 1,000,000 years ago. Perhaps if you stood in just the right place and yelled for their attention at just the right time . . .

. . . you could meet up for a snack.

57. AMMONIA TUBE

What would happen if you fed ammonia into your stomach through a tube? How fast must the flow rate be to burn your stomach from the heat released? What would the newly created chlorine gas do to your stomach?

—Becca

I'm a little concerned about your chemistry class.

This is definitely one of the more alarming questions I've gotten, but I have to admit I'm also extremely curious about the answer.

Derek Lowe, research chemist and author of the blog *In the Pipeline*, has a lot of firsthand experience with unpleasant chemicals, so I asked for his thoughts on what ammonia would do to the stomach. The good news, he told me, is that the reaction wouldn't produce chlorine gas. Ammonia is a base, so it would react directly with the acid in your stomach and neutralize it, forming a salt. The salt, ammonium chloride, is mildly irritating to your digestive system but not particularly harmful in itself. However, the above reaction also produces a lot of heat, so you'd suffer stomach burns as the acid and the ammonia neutralized.

Not all of the ammonia would be neutralized. "The limiting factor would be the acid," Lowe told me. There's not *that* much acid in your stomach, so it wouldn't take long for the ammonia to neutralize it all. "Then," he said, "you're on to direct tissue damage."

A review of ammonia toxicity, from the medical reference library StatPearls, includes the following phrases:

- "Inflammatory response"
- "Irreversible scarring"
- "Significant thermal injury"
- "Liquefaction necrosis"
- "Injuries along the alimentary canal"
- "Protein denaturation"
- "Perforation of the hollow viscera"
- "Saponification"

Saponification, if you're wondering, is the conversion of lipids—in this case, the membranes holding your cells together—to soap. That makes the inside of your cells fall out, which is bad for reasons that I *really* hope don't require an explanation.

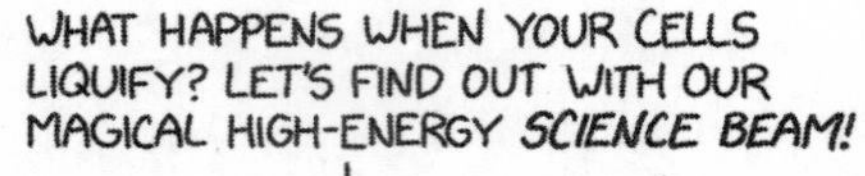

In conclusion:

1. Don't fill your stomach with ammonia.
2. Someone should probably check on Rebecca's chemistry class.

58. EARTH-MOON FIRE POLE

My son (5 years old) asked me today: If there were a kind of a firemen's pole from the Moon down to the Earth, how long would it take to slide all the way from the Moon to the Earth?

—Ramon Schönborn, Germany

First, let's get a few things out of the way:

In real life, we can't put a metal pole between the Earth and the Moon.* The end of the pole near the Moon would be pulled toward the Moon by the Moon's gravity, and the rest of it would be pulled back down to the Earth by the Earth's gravity. The pole would be torn in half.

Another problem with this plan: The Earth's surface spins faster than the Moon goes around, so the end that dangled down to the Earth would break off if you tried to connect it to the ground:

* For one, someone at NASA would probably yell at us.

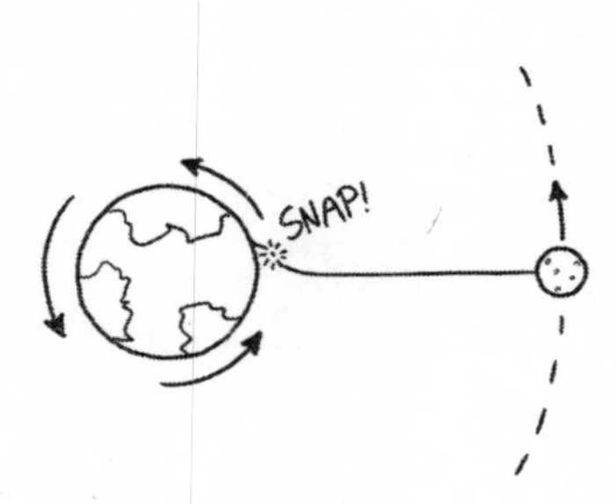

There's one* more problem: The Moon doesn't always stay the same distance from the Earth. Its orbit takes it closer and farther away. It's not a big difference, but it's enough that the thousands of kilometers of your fire-station pole would be squished against the Earth once a month.

But let's ignore those problems! What if we had a magical pole that dangled from the Moon down to just above the Earth's surface, expanding and contracting so it never quite touched the ground? How long would it take to slide down from the Moon?

If you stood next to the end of the pole on the Moon, a problem would become clear right away: You have to slide *up* the pole, and that's not how sliding works.

Instead of sliding, you'll have to climb.

People can climb poles pretty fast. World-record pole climbers† can climb at over a meter per second in championship competitions.‡ On the Moon, gravity is much

* Okay, that's a lie—there are, like, hundreds more problems.

† Of *course* there's a world record for pole climbing.

‡ Of *course* there are championship competitions.

weaker, so it will probably be easier to climb. On the other hand, you'll have to wear a spacesuit, so that will probably slow you down a little.

If you climb up the pole far enough, Earth's gravity will take over and start pulling you down. When you're hanging on to the pole, there are three forces pulling on you: The Earth's gravity pulling you toward Earth, the Moon's gravity pulling you away from Earth, and the centrifugal force from the swinging pole pulling you away from Earth.* At first, the combination of the Moon's gravity and the centrifugal force is stronger, pulling you toward the Moon, but as you get closer to the Earth, Earth's gravity takes over. The Earth is heavier than the Moon, so you'll reach this point—which is known as the L1 Lagrange point—while you're still pretty close to the Moon.

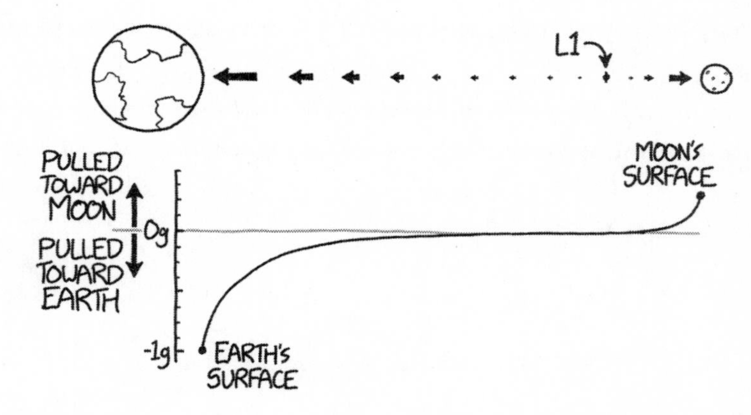

Unfortunately for you, space is big,[citation needed] so "pretty close" is still a long way. Even if you climb at better-than-world-record speed, it will still take you several years to get to the L1 crossover point.

As you approach the L1 point, you'll start to be able to switch from climbing to pushing-and-gliding: You can push once and then coast a long distance up the pole. You don't have to wait to stop, either—you can grab the pole again and give yourself a push to move even faster, like a skateboarder kicking several times to speed up.

* At the distance of the Moon's orbit and the speed it's traveling, centrifugal force pushing away is exactly balanced by the Earth's gravity—which is why the Moon orbits there.

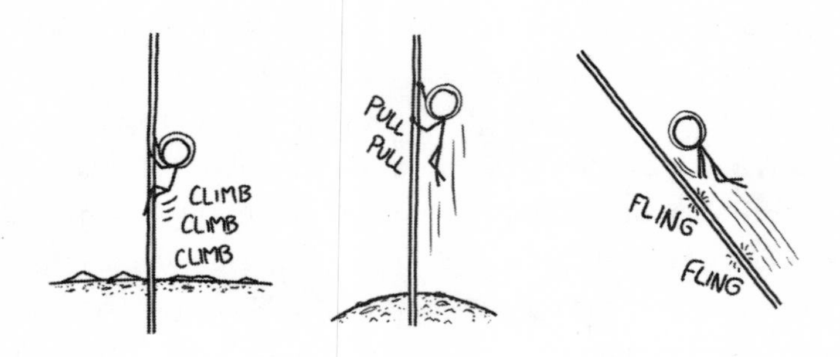

Eventually, as you reach the vicinity of the L1 point and are no longer fighting gravity, the only limit on your speed will be how quickly you can grab the pole and "throw" it past you. The best baseball pitchers can move their hands at about 100 mph while flinging objects past them, so you probably can't expect to move much faster than that.

Note: While you're flinging yourself along, be careful not to drift out of reach of the pole. Hopefully you brought some kind of safety line so that you can recover if that happens.

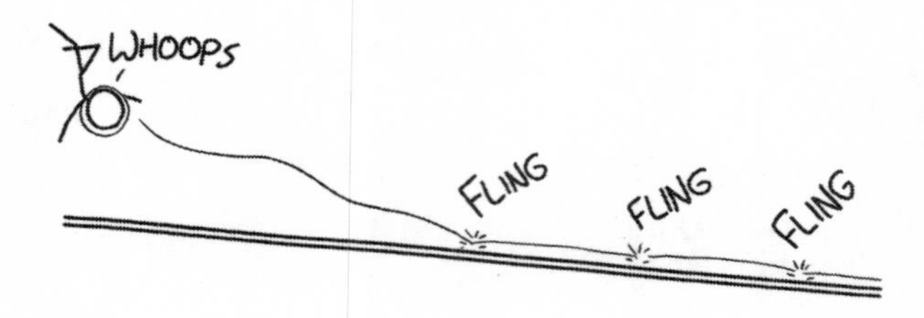

After another few weeks of gliding along the pole, you'll start to feel gravity take over, speeding you up faster than you can go by pushing yourself. When this happens, be careful—soon, you'll need to start worrying about going *too* fast.

As you approach the Earth and the pull of its gravity increases, you'll start to speed up quite a bit. If you don't stop yourself, you'll reach the top of the atmosphere at roughly escape velocity—11 km/s—and the impact with the air will produce so much heat that you risk burning up. Spacecraft deal with this problem using heat shields, which are capable of absorbing and dissipating this heat without burning up the spacecraft behind it. Since you have this handy metal pole, you can control your descent by clamping onto it and controlling your rate of descent through friction.

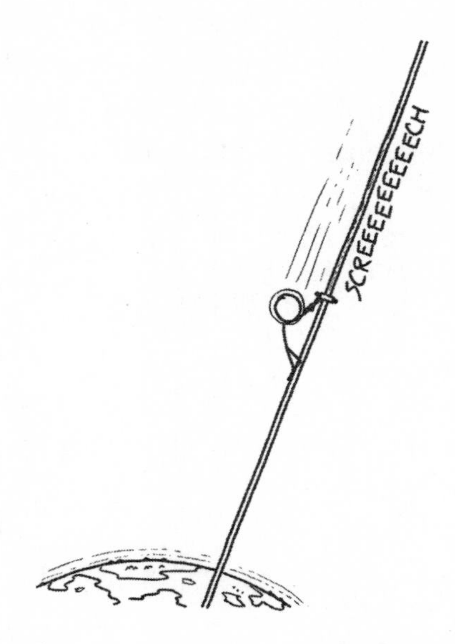

Be sure to keep your speed low during the whole approach and descent—and, if necessary, pause to let your hands or brake pads cool down—rather than wait until the end to try to slow down. If you get up to escape velocity, then at the last minute remember that you need to slow down, you'll be in for an unpleasant surprise as you try to grab on to the pole. At best, you'll be flung away and plummet to your death. At worst, your hands and the surface of the pole will both be converted into exciting new forms of matter, and *then* you'll be flung away and plummet to your death.

Assuming you descend slowly and enter the atmosphere in a controlled manner, you'll soon encounter your next problem: Your pole isn't moving at the same speed as the Earth. Not even close. The land and atmosphere below you are moving very fast relative to you. You're about to drop into some extremely strong winds.

The Moon orbits around the Earth at a speed of roughly 1 kilometer per second, making a wide loop every 29 days or so. That's how fast the top end of our hypothetical fire pole will be traveling. The bottom end of the pole makes a much smaller circle in the same amount of time, moving at an average speed of only about 35 mph relative to the center of the Moon's orbit.

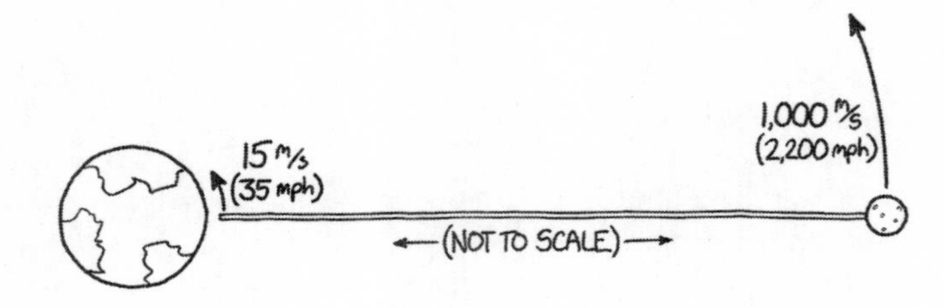

Now, 35 miles per hour doesn't sound bad. Unfortunately for you, the Earth is *also* spinning,* and its surface moves a *lot* faster than 35 mph; at the equator, it can reach over 1,000 miles per hour.†

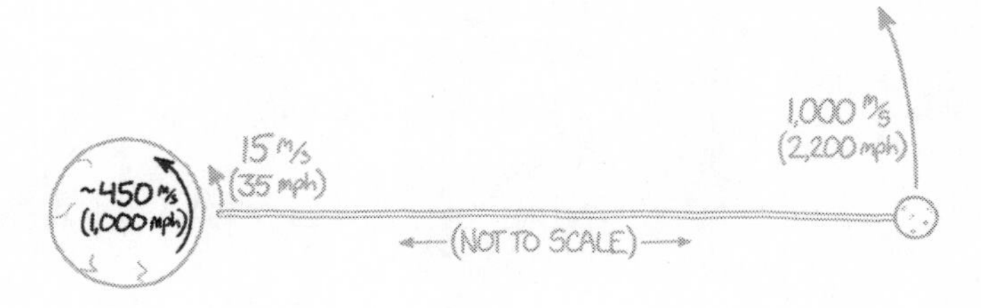

Even though the end of the pole is moving slowly relative to the Earth as a whole, it's moving very fast relative to the *surface*.

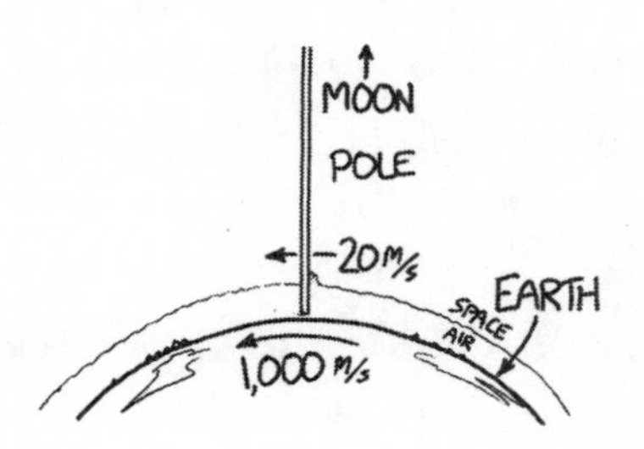

Asking how fast the pole is moving relative to the surface is effectively the same as asking what the ground speed of the Moon is. This is tricky to calculate, because the Moon's ground speed varies over time in a complicated way. Luckily for us, it doesn't

* I mean "unfortunately" in this specific context. In general, the fact that the Earth spins is very fortunate for you, and for the planet's overall habitability.

† It's common knowledge that Mount Everest is the tallest mountain on Earth, measured from sea level. A somewhat more obscure piece of trivia is that the point on the Earth's surface farthest from its center is the summit of Mount Chimborazo in Ecuador, due to the fact that the planet bulges out at the equator. Even *more* obscure is the question of which point on the Earth's surface moves the *fastest* as the Earth spins, which is the same as asking which point is farthest from the Earth's axis. The answer isn't Chimborazo *or* Everest. The fastest point turns out to be the peak of Mount Cayambe,‡ a volcano north of Chimborazo. You just learned that.

‡ Mt. Cayambe's southern slope also happens to be the highest point on Earth's surface directly on the equator. I have a lot of mountain facts.

vary *that* much—it's usually somewhere between 390 and 450 m/s, or a little over Mach 1—so figuring out the precise value isn't necessary.

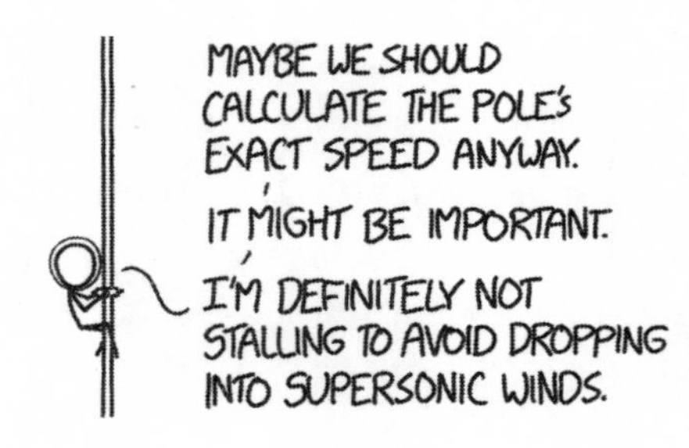

Let's buy a little time by trying to figure it out anyway.

The Moon's ground speed varies pretty regularly, making a kind of sine wave. It peaks twice every month as it passes over the fast-moving equator, then reaches a minimum when it's over the slower-moving tropics. Its orbital speed also changes depending on whether it's at the close or far point in its orbit. This leads to a roughly sine-wave-shaped ground speed:

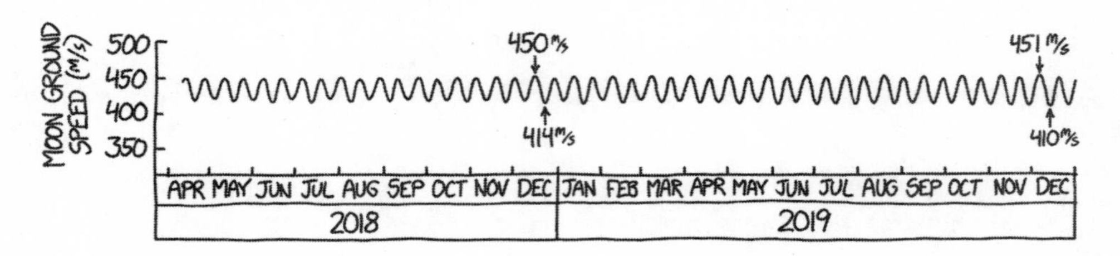

Well, ready to jump?

Okay, fine. There's one other cycle we can take into account to *really* nail down the Moon's ground speed. The Moon's orbit is tilted by about 5 degrees relative to the

Earth-Sun plane, while the Earth's axis is tilted by 23.5 degrees. This means that the Moon's latitude changes the way the Sun's does, moving from the northern tropics to the southern tropics twice a year.

However, the Moon's orbit is also tilted, and this tilt rotates on an 18.9-year cycle. When the Moon's tilt is in the same direction as the Earth's, it stays 5 degrees closer to the equator than the Sun, and when it's in the opposite direction, it reaches more extreme latitudes. When the Moon is over a point farther from the equator, it has a lower ground speed, so the lower end of the sine wave goes lower. Here's the plot of the Moon's ground speed over the next few decades:

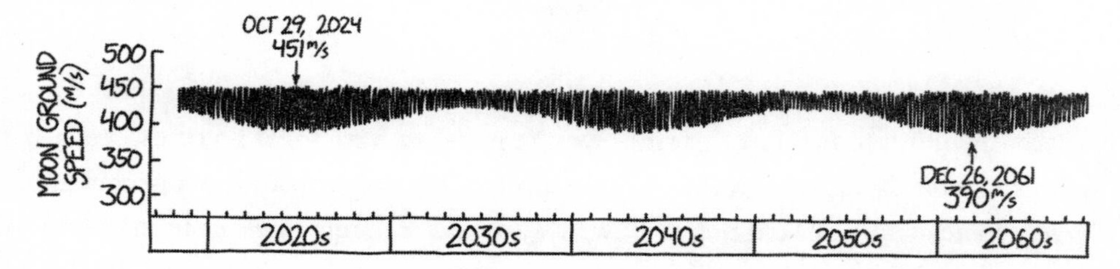

The Moon's top speed stays pretty constant, but the lowest speed rises and falls with an 18.9-year cycle. The lowest speed of the next cycle will be on May 1, 2025, so if you want to wait until 2025 to slide down, you can hit the atmosphere when the pole is moving at only 390 m/s relative to the Earth's surface.

When you do finally enter the atmosphere, you'll be coming down near the edge of the tropics. Try to avoid the tropical jet stream, an upper-level air current that blows in the same direction the Earth rotates. If your pole happens to go through it, it could add another 50 to 100 m/s to the wind speed.

Regardless of where you come down, you'll need to contend with supersonic winds, so you should wear lots of protective gear. Make sure you're tightly attached to the pole, since the wind and various shock waves will be violently battering and jolting you around. People often say, "It's not the fall that kills you, it's the sudden stop at the end." Unfortunately, in this case, it's probably going to be both.

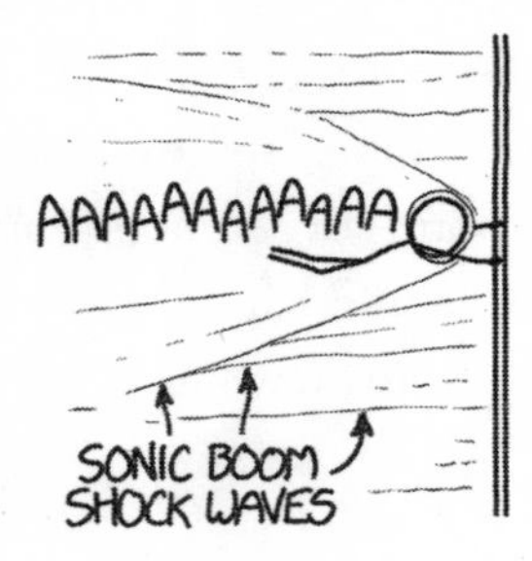

At some point, to reach the ground, you're going to have to let go of the pole. For obvious reasons, you don't want to jump directly onto the ground while moving at Mach 1. Instead, you should probably wait until you're somewhere near airline cruising altitude, where the air is still thin, so it's not pulling at you too hard—and let go of the pole. Then, as the air carries you away and you fall toward the Earth, you can open your parachute.

Then, at last, you can drift safely to the ground, having traveled from the Moon to the Earth completely under your own muscle power. Assuming you don't hang around the bottom of the pole for *too* long waiting to jump, the whole trip will take a few years—most of it spent shimmying up the pole near the Moon's surface.

When you're done, remember to remove the fire pole. That thing is *definitely* a safety hazard.

short answers #5

Q Could life evolve in a constantly running microwave?
—Abby Doth

WHAT THE ANSWER ACTUALLY IS ↓ / WHAT YOU'D EXPECT THE ANSWER TO BE →	YES	NO
YES	DOES MIT HAVE CLASSROOMS?	DOES AMHERST COLLEGE HAVE A NUCLEAR BUNKER?
NO	DO SCIENTISTS KNOW WHY LIGHTNING HAPPENS?	IS IT SAFE TO EAT RABID ANIMALS? CAN LIFE EVOLVE IN A RUNNING MICROWAVE?

Q Tonight at my work as an ER nurse in the emergency room, a patient (high on methamphetamine) asked for a cup of water. I returned with a paper cup of water, which the patient promptly threw at my head, missing me but hitting the wall in such an improbable way that the open top of the cup impacted the wall and the cup contained/diminished most of the subsequent spatter. It occurred to me that it might be possible to throw a cup of water hard enough that the container of water would go through the wall. Is this possible?

—Pete, RN

Sure, anything will go through a wall if you throw it hard enough. Also, I think this question might be a HIPAA violation.

Q How slow would you have to chew in order to be able to infinitely consume breadsticks?

—Miller Broughton

Olive Garden's breadstick "with garlic topping" is 140 calories, so in order to support your normal resting metabolism, you'd need to eat slightly less than one breadstick per hour.

If you divide each breadstick into 20 bites . . .

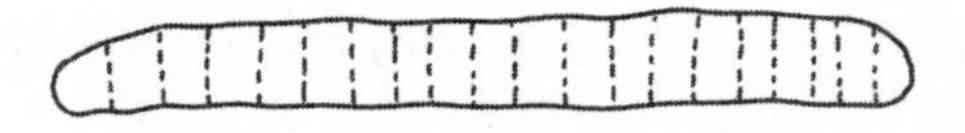

. . . and chew them at a rate of 1 second per chew . . .

. . . and chew each bite 200 times, which is twice as much as the 100 chews advocated by early-twentieth-century-chewing-obsessed weirdo Horace Fletcher, who was not a doctor . . .

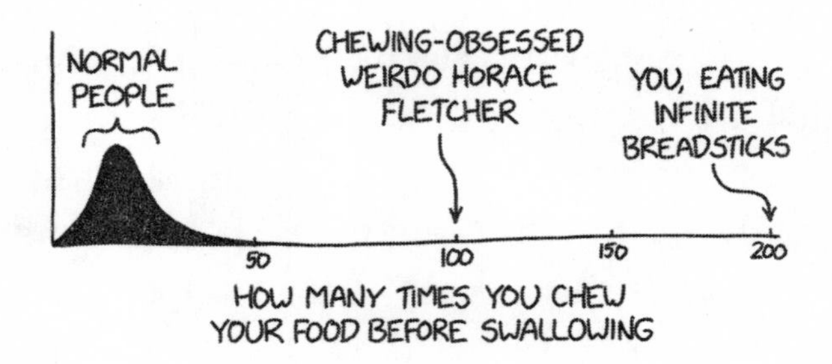

. . . then you can have infinite breadsticks.

Q If you were somehow to remove the white and yolk from inside an eggshell (chicken), and replace them with helium, would the eggshell float in the air?*

—Elizabeth

Nope! A medium egg is about 50 grams. But the air displaced by the shell weighs only about 50 milligrams, so even if it were filled with a vacuum, it wouldn't be able to lift more than 50 milligrams of weight.

---

* This question was inspired by an episode of the British competition show *Taskmaster*, in which contestant Mawaan Rizwan tried, unsuccessfully, to do just that.

An eggshell weighs a few grams, so it would stay on the ground.

There's a neat way to answer "Will it float?" questions without doing too many complicated calculations. Water is roughly 1,000 times denser than air,* so if you want to know whether something could float if you filled it with helium, just estimate how heavy it would be when filled with water, then move the decimal point over 3 places. That's how much buoyancy it could produce, so it's how light the solid parts have to be in order to float.

For example, a fish tank full of water might weigh 150 kilograms. That means that it displaces about 0.150 kilograms of air, or 150 grams, which is about the weight of a large smartphone. Since an empty fish tank definitely weighs more than a smartphone, a fish tank full of helium won't be able to float.

What would stars smell like, if it were possible to smell them?

—Finn Ellis

* The difference is actually more like 830 times, but if you round up to 1,000, it's both easier *and* just about perfectly compensates for the weight of the helium—which we were ignoring—to give you the correct answer. Sometimes, in calculations, two wrongs make a right!

Acrid and pungent, like bleach or burning rubber.

Stars are made of ionized plasma—a lot of charged particles whizzing about at high speed. There's no way to smell them without being burned. But let's imagine you took a sample of the plasma and slowed down the particles enough that you could take a whiff of it, without changing its chemical composition.

The plasma would immediately bind to the interior surface of your nose. Ionized particles are extremely chemically reactive, and the ions would start swapping electrons with your nasal lining and forming chemically reactive molecules—free radicals—in the mucus that covers your olfactory receptors. Those receptors are normally discerning, but these kinds of loose unbalanced molecules will bind to anything, so a lot of receptors would be triggered at once.

We can get an idea of what a star might smell like from a 1991 study that surveyed people whose nasal cavities had been irradiated during cancer treatment. They reported smelling an unpleasant odor when the machine was turned on, which they variously described as resembling "chlorine," "burning ammonia," "brakes burning," and "celery or bleach." The unpleasant smell from radiation treatments was likely caused by the gamma rays ionizing the mucus in their nasal lining and creating ozone and free radicals, activating their olfactory receptors in the same way that stellar plasma might.

In other words, stars probably don't smell great.

You can experience this smell yourself if you ever get a whiff of ozone, which is what creates that burning smell associated with electrical sparks. It's created by high-voltage equipment, some electric motors, and lightning strikes. But be careful not to breathe too much of it, since inhaling something that caustic isn't great for your nose, throat, or lungs.

It's actually much easier to guess what a star would taste like: sour. The sour receptors on our tongue are activated by free hydrogen ions, which we usually encounter in food in the form of acidic liquids. The bulk of a star's atmosphere is made of hydrogen ions, so it would very directly activate those receptors, giving a star an overwhelmingly sour taste.

Q What is the average size for every man-made object on the planet?

—Max Carver

Not too big, not too small. About average.

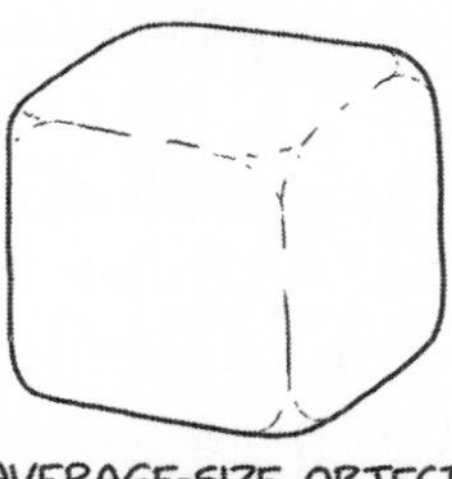

Q EEEEEEEEEEEEEEEEEEEEEEEEEEEEEEEEEEEEEEEEEEEE
EEEEEEEEEEEEEEEEEEEEEEEEEEEEEEEEEEEEEEEEEEEE
EEEEEEEEEEEEEEEEEEEEEEEEEEEEEEEEEEEEEEEEEEEE
EEEEEEEEEEEEEEEEEEEEEEEEEEEEEEEEEEEEEEEEEEEE
EEEEEEEEEEEEEEEEEEEEEEEEEEEEEEEEEEEEEEEEEEEE
EEEEEEEEEEEEEEEEEEEEEEEEEEEEEEEEEEEEEEEEEEEE
EEEEEEEEEEEEEEEEEEEEEEEEEEEEEEEEEEEEEEEEEEEE
EEEEEEEEEEEEEEEEEEEEEEEEEEEEEEEEEEEEEEEEEEEE
EEEEEEEEEEEEEEEEEEEEEEEEEEEEEEEEEEEEEEEEEEEE
EEEEEEEEEEEEEEEEEEEEEEEEEEEEEEEEEEEEEEEEEEEE
EEEEEEEEEEEEEEEEEEEEEEEEEEEEEEEEEEEEEEEEEEEE
EEEEEEEEEEEEEEEEEEEEEEEEEEEEEEEEEEEEEEEEEEEE
EEEEEEEEEEEEEEEEEEEEEEEEEEEEEEEEEEEEEEEEEEEE
EEEEEEEEEEEEEEEEEEEEEEEEEEEEEEEEEEEEEEEEEEEE
EEEEEEEEEEEEEEEEEEEEEEEEEEEEEEEEEEEEEEEEEEEE
EEEEEEEEEEEEEEEEEEEEEEEEEEEEEEEEEEEEEEEEEEEE
EEEEEEEEEEEEEEEEEEEEEEEEEEEEEEEEEEEEEEEEEEEE
EEEEEEEEEEEEEEEEEEEEEEEEEEEEEEEEEEEEEEEEEEEE
EEEEEEEEEEEEEEEEEEEEEEEEEEEEEEEEEEEEEEEEEEEE
EEEEEEEEEEEEEEEEEEEEEEEEEEEEEEEEEEEEEEEEEEEE
EEEEEEEEEEEEEEEEEEEEEEEEEEEEEEEEEEEEEEEEEEEE
EEEEEEEEEEEEEEE

—Nate Yu

59. GLOBAL SNOW

From my 7-year-old son, Owen:
How many snowflakes would it take to cover the entire world in 6 feet of snow? (I don't know why 6 feet . . . but that's what he asked.)

—Jed Scott

Snow is fluffy because it has a lot of air in it. The same amount of water that makes an inch of rain would make a lot more than an inch of snow.

An inch of rain is usually equal to about a foot of snow, but it depends on what kind of snow it is. If the snow is light and fluffy, then an inch worth of rain could make over 20 inches of snow!

All the clouds in the world, combined, hold about 13 trillion tons of water. If all that water were spread out evenly and fell at the same time, it would cover the Earth with an inch of rain—or a foot of snow.

Most of the Earth is ocean. If we only made water fall on land, there would be enough for 3 or 4 inches of water. That's how much falls in a very big rainstorm.

So 3 or 4 inches of water should add up to 3 or 4 feet of snow, right?

Almost, but there's a problem. When snow piles up, the snow on the bottom gets squished. If a foot of snow falls, then another foot falls, the snow on the bottom gets squished, which means that the whole pile is shorter than 2 feet high.

If you leave the snow there, it will slowly get less and less deep as it settles down and compacts. This means that even if 6 feet of snow fell everywhere, it would only be 6 feet at first. Before long, it might be 5 feet. (This happens to humans, too. You get shorter throughout the day as your body compresses a little!)

This can make it hard to record exactly how much snow falls, and sometimes even weather experts have a hard time! If you wait until the end of a snowstorm to measure snow, maybe it will have all squished down, or some of the snow might have melted, so your measurement will be too small.

Instead of waiting until the end of the storm, you can measure the snow in parts. You let some snow fall, measure it, then clear it away and wait for more snow to fall.

You have to decide how much snow to clear away. If you wait too long, the snow might become too squished, but if you measure it too often, it will all be light and fluffy and you'll get a number that's way too high.

Believe it or not, the National Weather Service has written special guidelines for how often to clear away snow, so everyone can measure it the same way. They use a special snow-measuring board, which is probably just a regular piece of wood, but I like to imagine that they treat it like a precision instrument and store it in a special locked case until it's needed.

The official guidelines say that you should clear the snow-measuring board every 6 hours. A few years ago, there was a big snowstorm, and the Baltimore airport measured 28.6 inches of snow. That would have been a new record. But then the National Weather Service learned that the person measuring the snow had cleared the board every hour instead of every 6 hours. So they didn't know whether to count it as a record or not.

I didn't see what they ended up deciding, because 4 days later, *another* blizzard hit Baltimore and everyone suddenly had more important things to worry about. (Then there were more after that one. It was a snowy winter.)

Still, people have never seen a winter with 6 feet of snow across the entire world.* A snowfall like that would—to answer the original question—take a total of about 10^{23} snowflakes, give or take a few zeros. With that much snow, every one of the 70 million kids in the United States would be able to make enough snowballs to hit every *other* kid with a snowball 3 times over.

Or, if it was a hot summer where you lived when the global snowfall happened, you could just keep the snowballs for yourself.

MMMM
SO COLD

* Unless the Toba catastrophe theory from chapter 56 turned out to be true.

60. DOG OVERLOAD

Assuming 1 out of every 4 people has a 5-year-old dog, and the dog reproduces once every year, with 5 puppies, and the puppies start reproducing at 5 and stop at 15 and die at 20, how long would it take for the Earth to be flooded with puppies, assuming we have all the food, water, and oxygen to sustain them?

—Griffin

If a quarter of Earth's 8 billion humans had a dog, that would be 2 billion dogs, which is already an awful lot. No one is exactly sure how many dogs there are in the world currently, but most estimates are less than 2 billion.

The next year, those 2 billion dogs would have 10 billion puppies,* boosting the total to 12 billion. That's enough for the other three-quarters of the population to all get a puppy of their own.

Over the first 5 years, those 2 billion dogs would continue to have 10 billion puppies each year. By the end of the fifth year, every human on Earth would have an average of 6 or 7 dogs.

In the sixth year, the puppies from the first year would start having puppies of their own, and exponential growth would really kick in. That year, the number of dogs would more than double from 52 billion to 112 billion. The next year it would nearly double again. By the eleventh year, we would reach the *101 Dalmatians* point, at which there would be 101 dogs for every human, about 85 percent of them younger than 5.

* I'm assuming each dog produces 5 puppies, instead of each *pair*. Either they're pairing up and having 10 puppies (5 per parent) or they're all female and reproducing parthenogenetically by cloning.

At the *Dalmatian* point, the combined biomass of the dogs would rival that of all other animal life on Earth combined. After another few years, there would be 1,001 dogs per human, and the land would start to become crowded. If the dogs were spread out evenly over the Earth's surface, they would be spaced about 5 meters apart.

After 15 years, the initial dogs would reach age 20—or 140 in dog years—and succumb to old age, but their number would be so small compared with the global population of about 10 trillion dogs that their disappearance would represent a rounding error.

After 20 years, the dogs would be spaced barely a meter apart across all the land area of the Earth, leaving us humans barely enough room to squeeze awkwardly among them. But no matter where you were, you'd be able to reach out a hand and pat a dog, so that's something.

After 25 or 30 years, the dogs would be shoulder to shoulder, and would begin to stack. Thankfully, the scenario guarantees them food and water and a long life,* so we'll assume these are dogs that enjoy stacking and tolerate it happily. By 40 years, our skyscrapers would begin to disappear beneath the barking, happy ocean of fur.

Over the next decade, the dogpile would subsume the mountains and spill out into the oceans. At this point the growth rate would be steady, with the number of dogs increasing by a factor of about 1.6578 every year. The total dog population in a given year can be estimated by a simple exponential function.

* No one who's read the first *What If?* book wants another *Mole of Moles* situation.

By the 55-year mark, the dogs would have displaced the atmosphere and outweighed the Moon. And after 65 years, as their population reached 1 mol ($6.022\text{x}10^{23}$), they would outweigh the world itself. Earth would no longer be a planet with dogs on it, it would be a bunch of dogs who found a planet to play with.

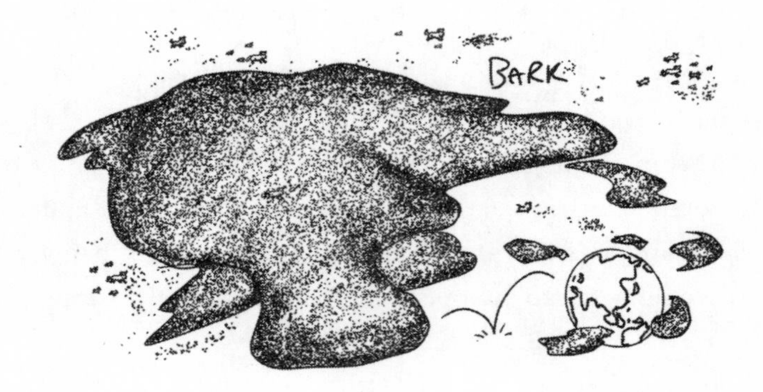

This can't go on forever. After 120 years, the outer edge of the expanding dog sphere would engulf the Sun. Even if we assume the dogs form some sort of a Dyson sphere to avoid this . . .

. . . after 110 years or so, as their population exceeded 10^{30}, they would exert a gravitational pull strong enough to undergo relativistic collapse.

If whatever force is keeping the dogs alive and happy also kept them from collapsing, we're so solidly outside the realm of physics that it doesn't even make sense to talk about what would happen. But for the record, here are the landmarks you'd hit:

- **150 years:** The dogs consume the Solar System, including the Kuiper belt
- **197 years:** The outer edge of the dog sphere starts expanding faster than the speed of light
- **200 years:** Dogs reach Sirius
- **250 years:** Dogs envelop the Milky Way
- **330 years:** The dog sphere encompasses the observable universe
- **417 years:** Disney releases their final film in the series

61. INTO THE SUN

When I was about 8 years old, shoveling snow on a freezing day in Colorado, I wished that I could be instantly transported to the surface of the Sun, just for a nanosecond, then instantly transported back. I figured this would be long enough to warm me up but not long enough to harm me. What would actually happen?

—AJ, Kansas City

Believe it or not, this wouldn't even warm you.

The temperature of the surface of the Sun is about 5,800 K,* give or take. If you stayed there for a while, you'd be cooked to a cinder, but a nanosecond is not very long—it's enough time for light to travel almost exactly a foot.†

* Or °C. When temperatures start having many digits in them, it doesn't really matter.

† A light-nanosecond is 11.8 inches (0.29979 meters), which is annoyingly close to a foot. I think it would be nice to redefine the foot as exactly 1 light nanosecond. This raises some obvious questions, like "Do we redefine the mile to keep it at 5,280 feet?" and "Do we redefine the inch?" and "Wait, *why* are we doing this?" But I figure other people can sort that out. I'm just the idea guy here.

I'm going to assume you're facing toward the Sun. In general, you should avoid looking directly at the Sun, but it's hard to avoid when it takes up a full 180 degrees of your view.

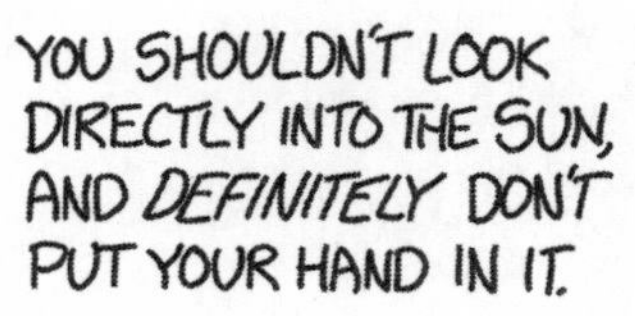

In that nanosecond, about a microjoule of energy would enter your eye.

A microjoule of light is not a lot. If you stare at a computer monitor with your eyes closed, then open and shut them quickly, your eye will take in about as much light from the screen during your reverse blink* as it would during a nanosecond on the Sun's surface.

* Is there a word for that? There should be a word for that.

During the nanosecond on the Sun, photons from the Sun would flood into your eye and strike your retinal cells. Then, at the end of the nanosecond, you'd jump back home. At this point, the retinal cells wouldn't even have begun responding. During the next few million nanoseconds (milliseconds) the retinal cells—having absorbed a bunch of light energy—would get into gear and start signaling your brain that something had happened.

You would spend 1 nanosecond on the Sun, but it would take 30,000,000 nanoseconds for your brain to notice. From your point of view, all you would see was a flash. The flash would seem to last much longer than your time on the Sun, only fading as your retinal cells quieted down.

The energy absorbed by your skin would be minor—about 10^{-5} joules per cm^2 of exposed skin. For comparison, according to the IEEE P1584 standard, holding your finger in the blue flame of a butane lighter for 1 second delivers about 5 joules per cm^2 to the skin, which is roughly the threshold for receiving a second-degree burn. The heat during your Sun visit would be 5 orders of magnitude weaker. Other than the dim flash in your eyes, you wouldn't even notice.

But what if you got the coordinates wrong?

The Sun's surface is relatively cool. It's hotter than, like, Phoenix,[citation needed] but compared to the interior, it's downright chilly. The surface is a few thousand degrees, but the interior is a few *million* degrees.* What if you spent a nanosecond *there*?

A PERSON WITHIN THE SUN
(NASA SIMULATION)

* The corona, the thin gas high above the surface, is *also* several million degrees, and no one knows why.

The Stefan-Boltzmann law lets us calculate how much heat you'd be exposed to while inside the Sun. It's not good. You would exceed the IEEE P1584B standard for second-degree burns after one *femto*second in the Sun. A nanosecond—the time you're spending there—is 1,000,000 femtoseconds. This does not end well for you.

There's some good news: Deep in the Sun, the photons carrying energy around have very short wavelengths—they're mostly a mix of what we'd consider hard and soft X-rays. This means they penetrate your body to various depths, heating your internal organs and also ionizing your DNA, causing irreversible damage before they even start burning you. Looking back, I notice that I started this paragraph with "there's some good news." I don't know why I did that.

In Greek legend, Icarus flew too close to the Sun and the heat melted his wings and he fell to his death. But "melting" is a phase change that is a function of temperature. Temperature is a measure of internal energy, which is the integral of incident power flux *over time*. His wings didn't melt because he flew too close to the Sun, they melted because he spent too much time there.

Visit briefly, in little hops, and you can go anywhere.

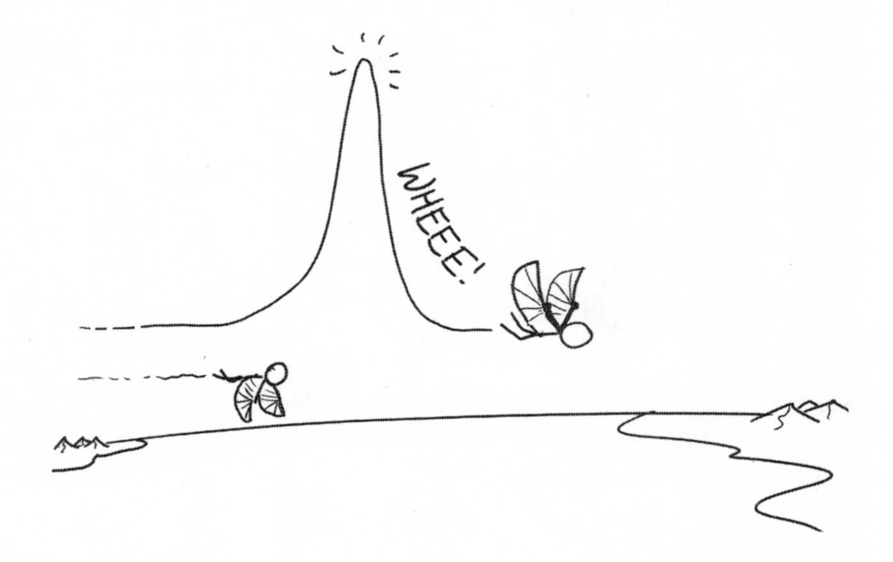

62. SUNSCREEN

Assuming that SPF works as it purports, what SPF would you need for a 1-hour trip to the surface of the Sun?

—Brian and Max Parker

When a sunscreen says SPF 20, it means that it should only let in 1/20th of the Sun's UV rays, allowing you to stay in the Sun 20 times longer before you get sunburned.

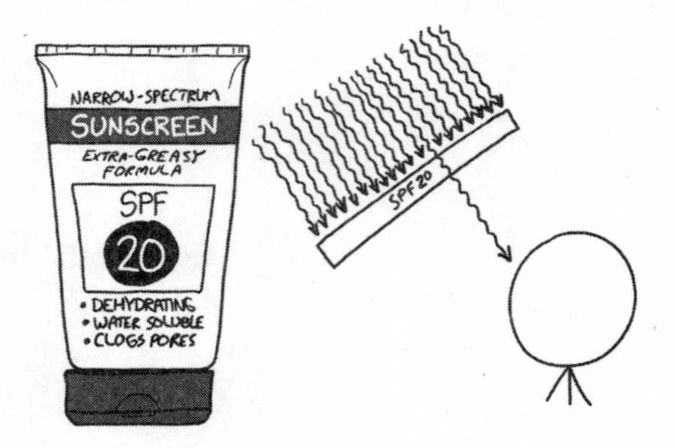

It's very hot close to the Sun.* Near the surface, the intensity of the heat and radiation is about 45,000 times greater than out here where the Earth orbits, so you would need SPF 45,000 just to cancel that out.

---

* Santana, C., I. Shur, R. Thomas, *Smooth* (New York, NY: Arista, 1999).

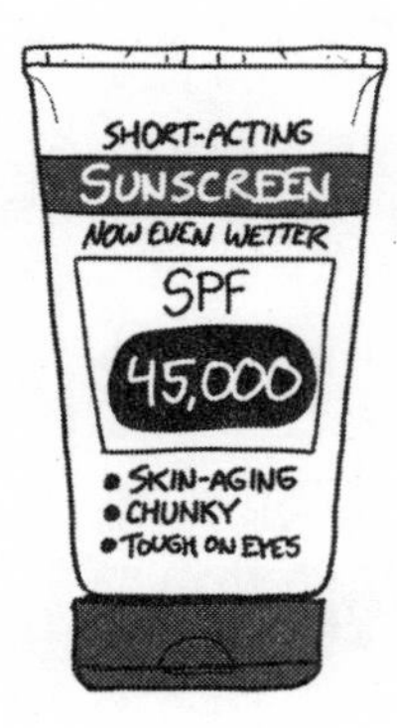

There's also more UV radiation in space in general, since you don't have the benefit of the Earth's atmosphere to protect you.

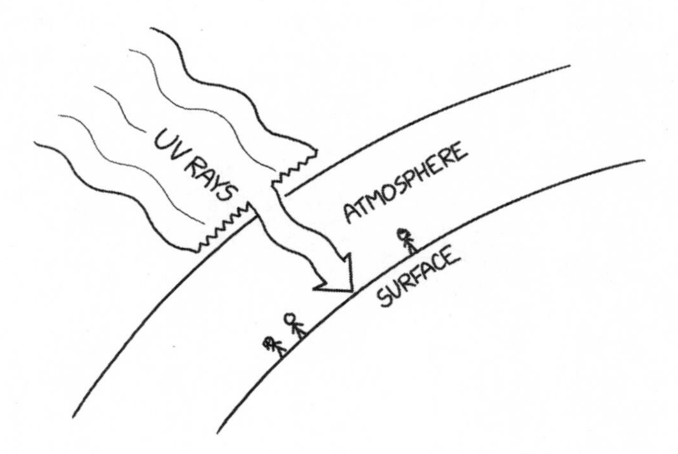

If astronauts didn't have UV-blocking suits, they'd sunburn much more quickly than on Earth. (There are stories that Apollo astronaut Gene Cernan tore enough layers of insulation in his spacesuit to get a bad sunburn on his lower back.)

The mix of wavelengths in space is a little different than it is on the surface, but the overall UV index in space might be about 30 times what it is on a sunny day on Earth. That means you'd need another 30-fold increase in protection, bringing the required SPF up to 1.3 million.

Luckily, that's not actually very much sunscreen! In theory, since SPF is a multiplier, when you put on several layers, you should multiply their SPF ratings together. If you put on one layer of SPF 20 sunscreen, then only 1/20th of the Sun's radiation should reach your skin. That means that if you put on a *second* layer of the same sunscreen, it should reduce that 1/20th by another 1/20th, for 1/400th total reduction. If that were true, 2 layers of SPF 20 sunscreen would be equivalent to SPF 400 sunscreen!

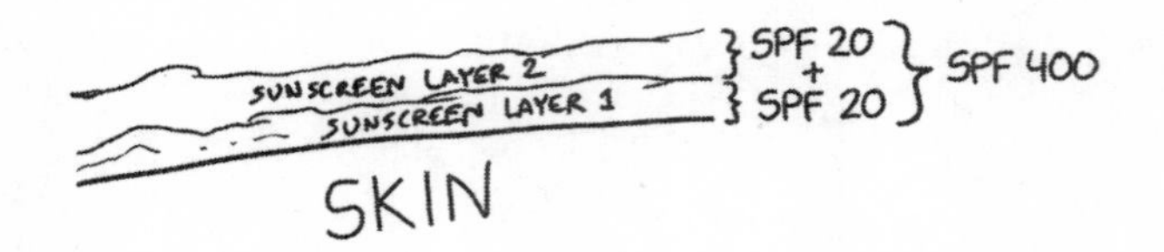

Five layers of SPF 20 sunscreen would be equivalent to SPF 3.2 million, enough to block the UV at the Sun's surface.

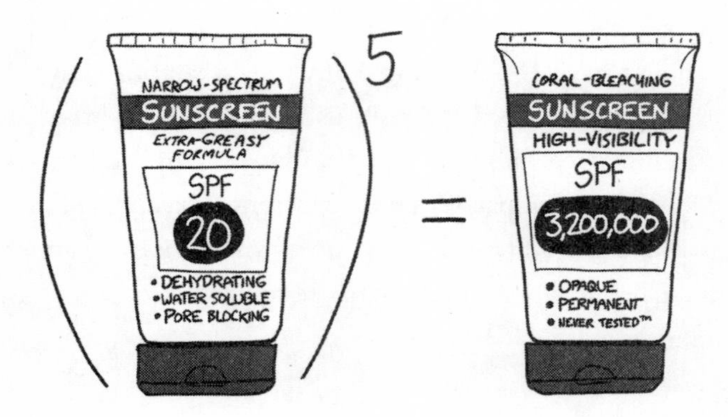

FDA testing standards say that sunscreen should be applied in a layer about 20 microns thick,* which means that in theory you'd need only 100 microns of SPF 20 sunscreen, about the thickness of a human hair, to keep you safe regardless of how close you get to the Sun.

This is obviously wrong for lots of reasons, but the biggest one is that sunscreen doesn't block the Sun's heat, just the UV rays. To successfully block the Sun's *heat* radiation, which is visible and infrared, you'd need a much thicker layer of sunscreen, which would itself heat up and boil away. Even sunscreen 10 meters thick wouldn't protect you from getting cooked.

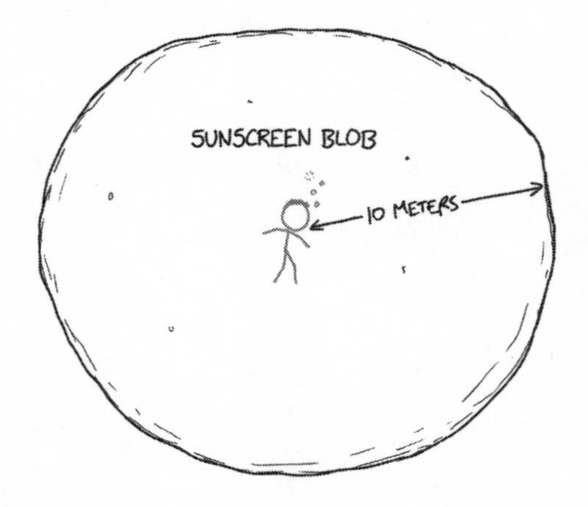

* In practice, sunscreen forms an irregular layer over the grooves and bumps of your skin, and most of the "sunburning" happens through the thinner "windows." Between the irregular layer and the fact that most people don't apply sunscreen thickly enough, SPF ratings are probably too high by a factor of 2 or more.

In theory, a large enough ball of sunscreen suspended near the Sun's surface could last long enough to protect you, but there's one other problem: You need to cover your whole body to avoid being vaporized, and it *clearly* says on the bottle to avoid getting it in your eyes.

We should probably add that to our list, too.

THINGS YOU SHOULD NOT DO
(PART 3,649 OF ????)

#156,824 EAT MEAT FROM RABID ANIMALS
#156,825 PERFORM YOUR OWN LASER EYE SURGERY
#156,826 TELL CALIFORNIA POULTRY REGULATORS THAT YOUR FARM IS SELLING POKEMON EGGS
#156,827 FUNNEL THE ENTIRE FLOW OF NIAGARA FALLS INTO THE OPEN WINDOW OF A PHYSICS LAB
#156,828 PUMP AMMONIA INTO YOUR ABDOMEN
#156,829 **(NEW!)** SUSPEND YOURSELF INSIDE A 10-METER BALL OF SUNSCREEN AND FALL INTO THE SUN

63. WALKING ON THE SUN

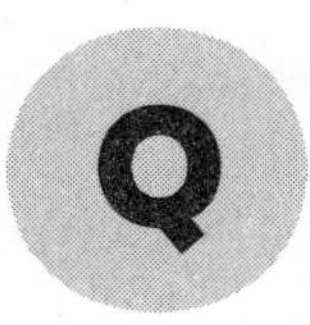

After the Sun runs out of fuel, it will become a white dwarf and slowly cool. When will it be cool enough to touch?

—Jabari Garland

The Sun will cool to room temperature in about 20 billion years.

Right now,* the Sun is getting hotter because the core is getting heavier, which makes its gravity pull harder and burn hydrogen faster. In about 5 billion years, it will start running out of hydrogen to burn. As the core collapses under its own weight, the heat of the collapse will trigger several desperate spasms of fusion that will inflate the outer layers† and then blast them away. Then what remains of the Sun will collapse to an inert, rapidly spinning ball slightly larger than the Earth—a white dwarf.

At first, the Sun's remnant will be white-hot from the violence of the collapse, but it will gradually cool over time as it radiates that heat into space. After a few billion years, it will be cooler than it is today. After 5 or 10 billion years, it will be the

* 2022

† And maybe consume the Earth.‡

‡ The fact that the destruction of the Earth is relegated to a footnote is a good sign for where this chapter is headed.

temperature of a campfire, radiating almost all of its heat in the infrared. Then, after another 10 or 20 billion years, it will reach room temperature.*

You can try to touch it, but you shouldn't. To see why, let's imagine you hop into a spaceship and fly toward it.

The Sun's white-dwarf remnant is much smaller than the old Sun. When your spaceship reaches the former location of the Sun's surface, the remnant Sun will only appear a little larger than the full Moon in the sky.†

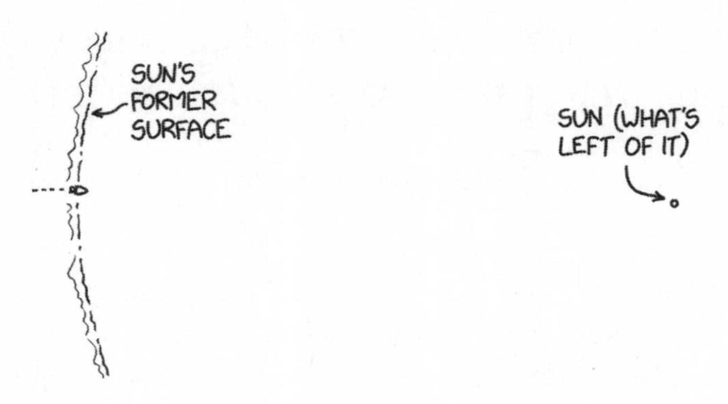

Unlike all the white dwarf stars that exist in the universe today, the Sun's remnant won't produce any light. You'll need headlights on your spaceship to see it.

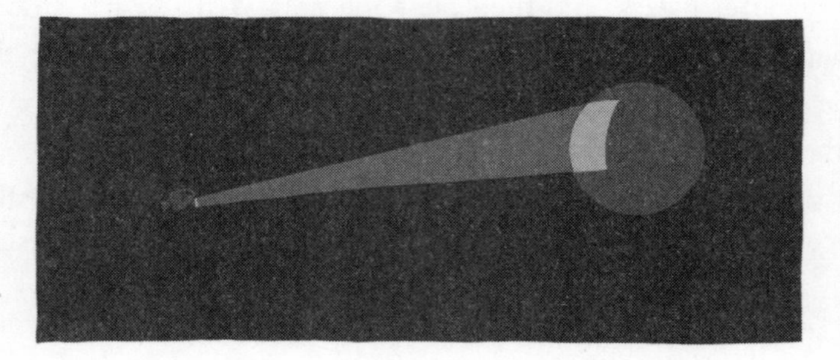

* There are no room-temperature stars in the sky right now because the universe isn't old enough. The first generation of white dwarf stars are still hot from their collapse. It will take many billions of years for them to cool down. The universe is still young.

† Back when we had a moon.‡

‡ And a sky

The surface will probably appear a dull gray color. Most of the atmosphere will have settled onto the surface under the immense pressure, but there may be a bluish haze from any hydrogen left over.

You'll feel fine while coasting toward the star, but if you try to stop your spaceship for a moment to admire the view, you'll run into trouble. The remnant still has about half the Sun's original mass, which means the gravitational pull at this distance will already be about 10 times Earth's gravity. If you try to hover in place or turn around, you'll black out from the g-forces unless you're wearing an acceleration suit.

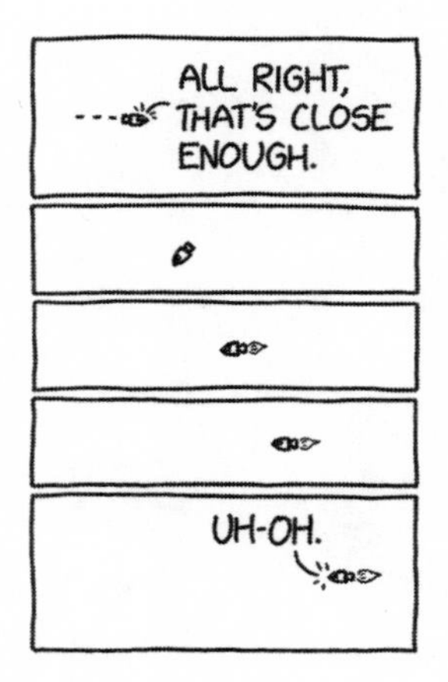

But if turning back is a mistake, continuing on is an even worse one, because there's no way to make a controlled landing on the surface of a cool dwarf star. It's not the fall that's the problem, it's the stop at the end. If you let yourself plunge toward the star, by the time you reached the surface you'd be moving at about 1 percent of the speed of light and would disintegrate on impact.

If you really want to land a vehicle on a white dwarf, you might try surfing. If you wait until the atmosphere has mostly settled onto the surface, you could send a vehicle into a surface-grazing orbit and try to slide along the surface to gradually slow yourself down. You'll need a giant ablative surfboard and you'll be riding on a layer of nuclear fusion. This is a bad plan and almost certainly wouldn't work, but I can't think of anything else you could try.

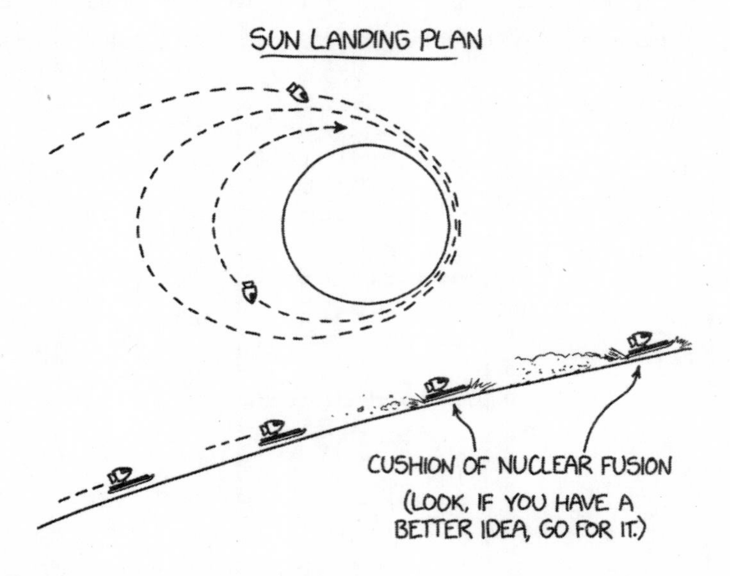

You'd need to send a robotic probe, since a human couldn't survive on the surface of a dwarf star; no pressure suit or support structure could keep you alive.

If you managed to set your robotic probe down gently on the surface of a stellar remnant, it wouldn't necessarily be crushed by gravity. A human couldn't survive there, but in theory some kind of computer might be able to. On a neutron star, which is much smaller and denser, any matter made of molecules is flattened out into a thin layer of atoms by the intense gravity, but on an Earth-size stellar remnant, some structures could support themselves.

On Earth, you can make small sculptures out of ice, but you can't form it into a mountain more than about a mile tall before it collapses under its own weight and flows like a glacier. On a stellar remnant, ice structures would be limited to about an *inch* of height. Other materials could support larger structures, but even a diamond—the hardest and most incompressible known substance—would crumble if formed into a skyscraper-size pyramid.

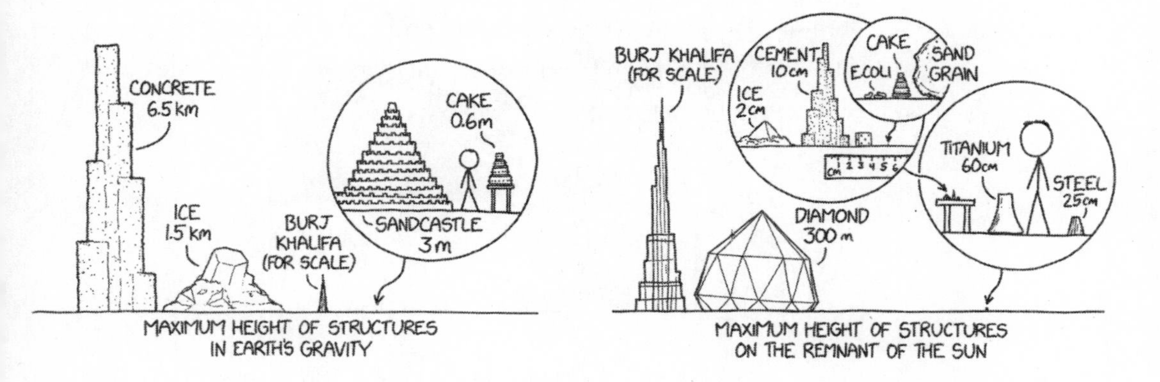

On Earth, a steel cable dangling from one end can be about 4 miles long before it snaps under its own weight. On a white dwarf star, cables could barely support 3 inches of their own weight. The largest suspension bridge on a dwarf star wouldn't be able to cross a gap more than an inch wide. Building a larger one would require high strength-to-weight materials like a spiderweb.

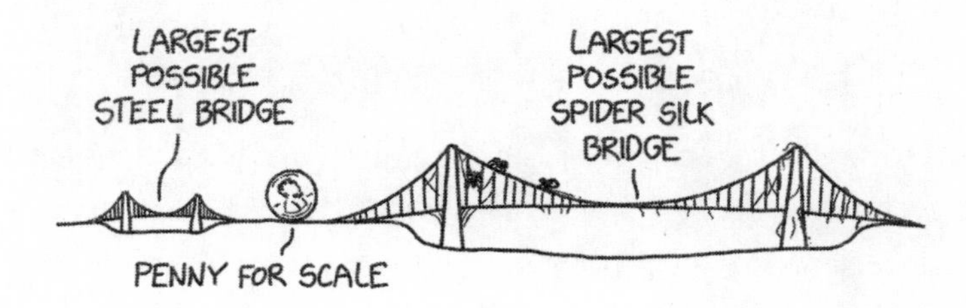

All this tells us that your lander will probably need to be ant-size rather than human-size, and you shouldn't count on having a lot of moving parts. But you might be able to build a small cube with some electronics embedded in it, capable of transmitting its observations back to you by radio.

Would landing a robotic probe count as *touching* the star? I don't know; that's sort of a philosophical question. But if you want to touch the star with your hand, then the answer is never. Even when a star cools to room temperature, there's no way to touch it with your own hand and survive.

And if you don't care about the survival part . . .

. . . then technically you could touch the Sun *now.*

64. LEMON DROPS AND GUMDROPS

What if all the raindrops were lemon drops and gumdrops?

—Shuo Peskoe-Yang

> *If all the raindrops were lemon drops and gumdrops*
> *Oh, what a rain that would be!*
> *I'd stand outside with my mouth open wide . . .*

—Children's song

This scenario is a catastrophe even by *What If* standards.

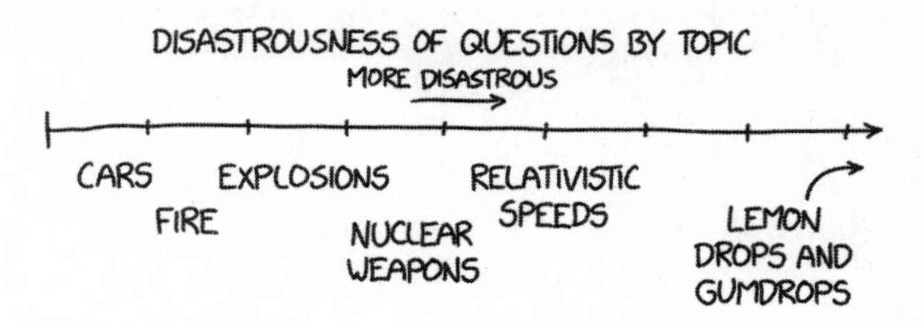

The terminal velocity of a lemon drop is about 10 meters per second. That's probably not fast enough to cause injury, but the lemon drops would definitely hurt as they bounced off your teeth.

Gumdrops are softer than lemon drops, so they wouldn't hurt quite as much, but catching them in your mouth still sounds like a good way to choke to death. You'd be better off just waiting for the storm to be over, then picking them up off the ground.

The first rain of lemon drops and gumdrops would be delicious. Once it was over, you could run through the fields, plucking candies from the ground and eating your fill, like excited children touring Willy Wonka's candy factory—with the one difference being that, on Wonka's factory tour, not *all* of the visitors died.

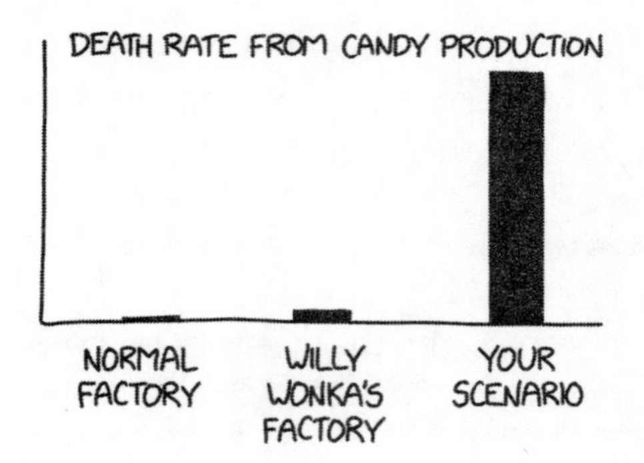

We'll assume the water is replaced by an equivalent mass of lemon drops and gumdrops, so a typical rainstorm would blanket the ground in ankle-deep candy. Unlike rainwater, the candies wouldn't soak into the soil or flow downhill. They'd just lie there on the ground. Children and animals would make a small dent in the pile, and sugar-digesting bacteria would bloom on others, but the bulk of the candy would just lie there to melt in the Sun.

After a few weeks of lemon-drop-gumdrop rain, the roof collapses would begin.

Home roofs in snowy areas are typically required to hold between 20 and 60 pounds of weight per square foot, which is the equivalent of about 30 centimeters of water. The eastern United States gets about a meter of rain every year, which means that within a few months most flat roofs would have collapsed under the weight.

We wouldn't all die of thirst right away. There's plenty of water in aquifers and lakes to sustain us for quite a while, although the surface water would become increasingly high-calorie.

Agriculture would collapse. The abrupt end of water-based rain would cause an immediate global drought. Many crops are watered by irrigation systems that rely on lakes and aquifers, but even those would be quickly buried under piles of candy. If your crops did manage to survive, harvesting them would be a nightmare—good luck driving your tractor through a sticky layer of knee-deep lemon drops and gumdrops.

Within a few years, most human cities would be buried beneath blankets of sugar, a whole planet of Candyland Pompeiis.

The places where agriculture would survive the longest would be desert areas where crops are watered almost entirely by irrigation, like the farmlands along the Nile River in Egypt, California's Imperial Valley, or the deserts of Turkmenistan. Cities like Cairo and Lima, which get virtually no annual rainfall, would be able to continue a relatively candy-free existence for years, although the destruction of the rest of the world would cause some problems.

Ultimately, our species would be unlikely to survive for long, but the consequences of the lemon drops and gumdrops scenario would be a lot worse than simple human extinction. Within just a few days, the candy would outweigh all living things on Earth, and adding such a massive blanket of sugar to the Earth would fundamentally reshape the planet.

Sugar is a carbohydrate, and it can be decomposed into CO_2 and water. This releases energy, which is why sugar is so popular with energetic living things like children, hummingbirds, and bacteria. If you add sugar to the soil, much of it will be digested by bacteria and returned to the environment in the form of CO_2 and water.

Anything that can live off of sugar would suddenly find itself in an environment with no limits. A lot of the candy would be buried undigested, but some of it would be digested or oxidized by other processes—like fire. When that happened, CO_2 levels would skyrocket and the planet would heat up.

Lemon drops and gumdrops are denser than water,* so the ones that fell in the ocean would sink before they dissolved, leaving the sea surfaces exposed to the atmosphere. As the planet warmed, water would evaporate faster and faster from the surface of the hot and increasingly sugary ocean.

* Citation: I just poured a glass of water and tried dropping various candies in. Science!

If a planet with oceans gets too hot, the atmosphere can fill with water vapor. This water vapor can trap more heat, leading to a feedback loop of out-of-control warming that continues until the oceans boil away. Something like this may have happened to Venus in the distant past. Luckily, after some nerve-racking calculations, scientists have generally concluded that the Earth isn't in danger of a runaway greenhouse effect anytime soon. There's just not enough CO_2 in the atmosphere to trigger an ocean-boiling heat spiral, even if we burn every last little bit of fossil fuel on the planet.

But the candy could do it. If even a fraction of the carbon in the candy was oxidized, it would push atmospheric CO_2 levels up from their current* 0.042 percent to 5 percent or 10 percent within a few years, levels not seen since the Earth was young

* I estimate this stat will become incorrect in roughly December 2024.

and the Sun was cooler and smaller. These levels, models suggest, might be able to trigger a runaway greenhouse effect.

Global temperatures would rise to furnace-like levels, virtually sterilizing the planet's surface and bringing an end to the tree of life. Save perhaps for some lucky sugar-eating thermophilic bacteria, no life would remain to watch as the planet's water boiled away. Soon, Earth would be a scorched, lifeless rock, with the ocean floors coated in the sugary gunk left behind when the sugar-choked oceans boiled away.

A final silver lining: Once the oceans boiled away, there would be no more raindrops to turn into lemon drops and gumdrops, so at least that rain would end. Earth might look a lot like Venus, with little to no water vapor and temperatures too hot for it to condense into rain.

Venus isn't completely free of precipitation. Its mountaintops are coated with a substance we call "snow"—really more like frost—which appears to be metal evaporated from the lowlands and deposited on the mountains. On a post-runaway-greenhouse Earth, we might be like Venus, with our dry, scorched mountaintops dusted with metallic snow.

Maybe we should just skip the next verse.

Acknowledgments

A lot of people helped make this book possible.

Thank you to everyone who generously shared their expertise with me. Thank you to Cindy Keeler for answering my questions about high-energy particles; Derek Lowe for insight into ammonia and triboluminescence; and Natalie Mahowald for telling me not to breathe iron vapor. Thank you to A. J. Blechner, Jonathan Zittrain, Jack Cushman, and everyone at Harvard's Library Innovation Lab for answering my questions about laws; Katie Mack for answering questions about space and time; and Maya Bergamasco of Harvard and Derek Spelay of the International Joint Commission for providing information on the mysterious and secretive international waterfall police who guard Niagara Falls. Thank you to Phil Plait for answering questions about telescopes, and Tracy Wilson for weighing whipped cream. And thank you to the federal prosecutor who told me that committing crimes is bad but asked to remain anonymous because "it's funnier that way."

Thank you to Kat Hagan, Janelle Shane, Reuven Lazarus, and Nick Murdoch for reading my answers and giving commentary; and to Christopher Night for the astonishing project of fact-checking this book, running the numbers on everything from structure size limits on a white dwarf star to the number of mushrooms in *Mario* levels. Any errors that remain are mine.

Thank you to my editor, Courtney Young, for believing in me from the start and for shepherding this book through to publication; and to the whole team at Riverhead, including Lorie Young, Jenny Moles, Kim Daly, Ashley Sutton, Ashley Garland, Jynne Martin, Geoff Kloske, Gabriel Levinson, Melissa Solis, Caitlin Noonan, Claire Vaccaro, Helen Yentus, Grace Han, Tyriq Moore, Linda Friedner, and Anna Scheithauer.

Thank you to Christina Gleason, incredibly talented designer and dear friend, who assembled my words into the shape of a book. Thank you to Casey Blair for overseeing the whole project and heroically keeping everything on track. Thank you to Marissa Gunning for organizational help, and to Derek for helping make the whole thing happen, and to my agent Seth Fishman and the folks at The Gernert Co., including Jack Gernert, Rebecca Gardner, Will Roberts, and Nora Gonzalez.

Thank you to everyone who sent in their questions. Thank you to the researchers whose work made it possible to answer them. And thank you to my wife—for being curious about everything, excited about the world, and for always looking for an adventure.

References

1. Soupiter

Lewis, Geraint F., and Juliana Kwan, "No Way Back: Maximizing Survival Time Below the Schwarzschild Event Horizon," *Publications of the Astronomical Society of Australia*, 2007, https://arxiv.org/abs/0705.1029.

2. Helicopter Ride

Anthony, Julian, and Wagdi G. Habashi, "Helicopter Rotor Ice Shedding and Trajectory Analyses in Forward Flight," *Journal of Aircraft* 58, no. 5 (April 28, 2021), https://doi.org/10.2514/1.C036043.

Liard, F. (ed.), *Helicopter Fatigue Design Guide*, Advisory Group for Aerospace Research and Development, November 1983, https://apps.dtic.mil/dtic/tr/fulltext/u2/a138963.pdf.

3. Dangerously Cold

O'Connor, BS, Mackenzie, Jordan V. Wang, MD, MBE, MBA, and Anthony A. Gaspari, MD, "Cold Burn Injury After Treatment at Whole-Body Cryotherapy Facility," *JAAD Case Reports* 5, no. 1 (December 4, 2018): 29–30, https://www.ncbi.nlm.nih.gov/pmc/articles/PMC6280691/.

Raman, Aaswath P., Marc Abou Anoma, Linxiao, Eden Raphaeli, and Shanhui Fan, "Passive Radiative Cooling Below Ambient Air Temperature Under Direct Sunlight," *Nature* 515 (2014): 540–544, https://doi.org/10.1038/nature13883.

"Safe Handling of Cryogenic Liquids," *Health & Safety Manual: Section 7: Safety Guidelines & SOP's*, University of California, Berkeley: College of Chemistry, https://chemistry.berkeley.edu/research-safety/manual/section-7/cryogenic-liquids.

"Safety Instructions: Cryogenics Liquid Safety," Oregon State University: Environmental Health & Safety, https://ehs.oregonstate.edu/sites/ehs.oregonstate.edu/files/pdf/si/cryogenics_si.pdf.

Sun, Xingshu, Yubo Sun, Zhiguang Zhou, Muhammad Ashraful Alam, and Peter Bermel, "Radiative Sky Cooling: Fundamental Physics, Materials, Structures, and Applications," *Nanophotonics* 6, no. 5 (July 29, 2017): 997–1015, https://www.degruyter.com/document/doi/10.1515/nanoph-2017-0020/html.

4. Ironic Vaporization

"Iron (Fe) Pellets Evaporation Materials," Kurt J. Lesker Company, https://www.lesker.com/newweb/deposition_materials/depositionmaterials_evaporationmaterials_1.cfm?pgid=fe1.

Mahowald, Natalie M., Sebastian Engelstaedter, Chao Luo, Andrea Sealy, Paulo Artaxo, Claudia Benitez-Nelson, Sophie Bonnet, Ying Chen, Patrick Y. Chuang, David D. Cohen, Francois Dulac, Barak Herut, Anne M. Johansen, Nilgun Kubilay, Remi Losno, Willy Maenhaut, Adina Paytan, Joseph M. Prospero, Lindsey M. Shank, and Ronald L. Siefert, "Atmospheric Iron Deposition: Global Distribution, Variability, and Human Perturbations," *Annual Review of Marine Science* 1 (January 2009): 245–278, https://www.annualreviews.org/doi/abs/10.1146/annurev.marine.010908.163727.

Spalvins, T., and W. A. Brainard, "Ion Plating with an Induction Heating Source," NASA Lewis Research Center, January 1, 1976, https://ntrs.nasa.gov/citations/19760010307.

5. Cosmic Road Trip

"Early Estimate of Motor Vehicle Traffic Fatalities for the First Quarter of 2021," *Traffic Safety Facts*, National Highway Traffic Safety Administration, U.S. Department of Transportation, August 2021, https://www.nhtsa.gov/sites/nhtsa.gov/files/2021-09/Early-Estimate-Motor-Vehicle-Traffic-Fatalities-Q1-2021.pdf.

"NHTSA Releases Q1 2021 Fatality Estimates, New Edition of 'Countermeasures That Work,'" National Highway Traffic Safety Administration, U.S.

Department of Transportation, September 2, 2021, https://www.nhtsa.gov/press-releases/q1-2021-fatality-estimates-10th-countermeasures-that-work.

6. Pigeon Chair

Abs, Michael, *Physiology and Behaviour of the Pigeon* (Cambridge, MA: Academic Press, 1983), 119.

Berg, Angela M., and Andrew A. Biewener, "Wing and Body Kinematics of Takeoff and Landing Flight in the Pigeon (*Columba livia*)," *Journal of Experimental Biology* 213 (May 15, 2010): 1651–1658, https://journals.biologists.com/jeb/article/213/10/1651/9685/Wing-and-body-kinematics-of-takeoff-and-landing.

Callaghan, Corey T., Shinichi Nakagawa, and William K. Cornwell, "Global Abundance Estimates for 9,700 Bird Species," *Proceedings of the National Academy of Sciences of the United States of America*, May 25, 2021, https://www.pnas.org/content/118/21/e2023170118/tab-figures-data.

Liu, Ting Ting, Lei Cai, Hao Wang, Zhen Dong Dai, and Wen Bo Wang, "The Bearing Capacity and the Rational Loading Mode of Pigeon During Takeoff," *Applied Mechanics and Materials* 461 (November 2013): 122–127, https://www.scientific.net/AMM.461.122.

Pennycuick, C. J., and G. A. Parker, "Structural Limitations on the Power Output of the Pigeon's Flight Muscles," *Journal of Experimental Biology* 45, (December 1, 1966): 489–498, https://journals.biologists.com/jeb/article/45/3/489/34321/Structural-Limitations-on-the-Power-Output-of-the.

Ⓢ Short Answers #1

Bates, S. C., and T. L. Altshuler, "Shear Strength Testing of Solid Oxygen," *Cryogenics* 35, no. 9 (September 1995): 559–566, https://www.sciencedirect.com/science/article/abs/pii/001122759591254I.

7. T. Rex Calories

Barrick, Reese E., and William J. Showers, "Thermophysiology and Biology of Gigantosaurus: Comparison with Tyrannosaurus," *Palaeontologia Electronica* 2, no. 2 (1999), https://web.archive.org/web/20210612062144/https://palaeo-electronica.org/1999_2/gigan/issue2_99.htm.

Hutchinson, John R., Karl T. Bates, Julia Molnar, Vivian Allen, and Peter J. Makovicky, "A Computational Analysis of Limb and Body Dimensions in Tyrannosaurus rex with Implications for Locomotion, Ontogeny, and Growth," *PLOS ONE* 9, no. 5 (2011), https://journals.plos.org/plosone/article?id=10.1371/journal.pone.0026037.

McNab, Brian K., "Resources and Energetics Determined Dinosaur Maximal Size," *PNAS* 106, no. 29 (2009): 12184–12188, https://www.pnas.org/content/106/29/12184.full.

O'Connor, Michael P., and Peter Dodson, "Biophysical Constraints on the Thermal Ecology of Dinosaurs," *Paleobiology* 25, no. 3 (1999): 341–368, https://www.jstor.org/stable/2666002.

8. Geyser

Hutchinson, Roderick A., James A. Westphal, and Susan W. Kieffer, "In Situ Observations of Old Faithful Geyser," *Geology* 25, no. 10 (1997): 875–878, https://doi.org/10.1130/0091-7613(1997)025<0875:ISOOOF>2.3.CO;2.

Karlstrom, Leif, Shaul Hurwitz, Robert Sohn, Jean Vandemeulebrouck, Fred Murphy, Maxwell L. Rudolph, Malcolm J. S. Johnston, Michael Manga, and R. Blaine McCleskey, "Eruptions at Lone Star Geyser, Yellowstone National Park, USA: 1. Energetics and Eruption Dynamics," *Journal of Geophysical Research: Solid Earth* 118, no. 8 (June 19, 2013): 4048–4062, https://agupubs.onlinelibrary.wiley.com/doi/abs/10.1002/jgrb.50251.

Kieffer, Susan, "Geologic Nozzles," *Reviews of Geophysics* 27, no. 1 (February 1989): 3–38, http://seismo.berkeley.edu/~manga/kieffer1989.pdf.

O'Hara, D. Kieran, and E. K. Esawi, "Model for the Eruption of the Old Faithful Geyser, Yellowstone National Park," *GSA Today* 23, no. 6 (June 2013): 4–9, https://www.geosociety.org/gsatoday/archive/23/6/article/i1052-5173-23-6-4.htm.

"Superintendents of the Yellowstone National Parks Monthly Reports, June 1927," Yellowstone National Park, 1927, https://archive.org/details/superintendents027june.

Whittlesey, Lee H., *Death in Yellowstone: Accidents and Foolhardiness in the First National Park* (Plymouth, England: Roberts Rinehart Publishers, 1995).

10. Reading Every Book

Buringh, Eltjo, and Jan Luiten Van Zanden, "Charting the 'Rise of the West': Manuscripts and Printed Books in Europe, A Long-Term Perspective from the Sixth through Eighteenth Centuries," *The Journal of Economic History* 69, no. 2 (2009): 409–445. doi:10.1017/S0022050709000837.

Grout, James, "The Great Library of Alexandria," *Encyclopaedia Romana*, http://penelope.uchicago.edu/~grout/encyclopaedia_romana/greece/paganism/library.html.

Pelli, Denis, and C. Bigelow, "A Writing Revolution," *Seed: Science Is Culture* (2009), https://web.archive.org/web/20120331052409/http://seedmagazine.com/supplementary/a_writing_revolution/pelli_bigelow_sources.pdf.

11. Banana Church

Grant, Amy, "Banana Tree Harvesting: Learn How and When to Pick Bananas," *Gardening Know How*, https://www.gardeningknowhow.com/edible/fruits/banana/banana-tree-harvesting.htm.

Pew Research Center, "How Religious Commitment Varies by Country Among People of All Ages," *The Age Gap in Religion Around the World*, June 13, 2018, https://www.pewforum.org/2018/06/13/how-religious-commitment-varies-by-country-among-people-of-all-ages/.

Stark Bro's., "Harvesting Banana Plants," The Growing Guide: How to Grow Banana Plants, https://www.starkbros.com/growing-guide/how-to-grow/fruit-trees/banana-plants/harvesting.

12. Catch!

Centers for Disease Control and Prevention, "Morbidity and Mortality Weekly Report," *MMWR* 53, no. 50 (December 24, 2004), https://www.cdc.gov/mmwr/PDF/wk/mm5350.pdf.

Close Focus Research, "Maximum Altitude for Bullets Fired Vertically," http://www.closefocusresearch.com/maximum-altitude-bullets-fired-vertically.

"Model 1873 U.S. Springfield at Long Range," *Rifle Magazine* 35, no. 5 (2003), https://web.archive.org/web/20160409042559/https://www.riflemagazine.com/magazine/article.cfm?magid=78&tocid=1094.

13. Lose Weight the Slow and Incredibly Difficult Way

Blackwell, David, Maria Richards, Zachary Frone, Joe Batir, Andrés Ruzo, Ryan Dingwall, and Mitchell Williams, "Temperature-at-Depth Maps for the Conterminous US and Geothermal Resource Estimates," SMU Geothermal Lab, October 24, 2011, https://www.smu.edu/Dedman/Academics/Departments/Earth-Sciences/Research/GeothermalLab/DataMaps/TemperatureMaps.

16. Star Sand

Abuodha, J. O. Z., "Grain Size Distribution and Composition of Modern Dune and Beach Sediments, Malindi Bay Coast, Kenya," *Journal of African Earth Sciences* 36 (2003): 41–54, http://www.vliz.be/imisdocs/publications/37337.pdf.

Stauble, Donald K., "A Review of the Role of Grain Size in Beach Nourishment Projects," U.S. Army Engineer Research and Development Center: Coastal and Hydraulics Laboratory, 2005, https://www.fsbpa.com/05Proceedings/02-Don%20Stauble.pdf.

17. Swing Set

Case, William B., and Mark A. Swanson, "The Pumping of a Swing from the Seated Position," *American Journal of Physics* 58, no. 463 (1990), https://aapt.scitation.org/doi/10.1119/1.16477.

Curry, Stephen M., "How Children Swing," *American Journal of Physics* 44, no. 924 (1976), https://aapt.scitation.org/doi/10.1119/1.10230.

Post, A. A., G. de Groot, A. Daffertshofer, and P. J. Beek, "Pumping a Playground Wing," *Motor Control* 11, no. 2 (2007): 136–150, https://research.vu.nl/en/publications/pumping-a-playground-swing.

Roura, P., and J. A. González, "Towards a More Realistic Description of Swing Pumping Due to the Exchange of Angular Momentum," *European Journal of Physics*

31, no. 5 (August 3, 2010), https://iopscience.iop.org/article/10.1088/0143-0807/31/5/020.

Wirkus, Stephen, Richard Rand, and Andy Ruina, "How to Pump a Swing," *The College Mathematics Journal* 29, no. 4 (2018): 266–275, https://www.tandfonline.com/doi/abs/10.1080/07468342.1998.11973953.

18. Airliner Catapult

"Eco-Climb," Airbus, https://web.archive.org/web/20170111010030/https://www.airbus.com/innovation/future-by-airbus/smarter-skies/aircraft-take-off-in-continuous-eco-climb/.

Chati, Yashovardhan S., and Hamsa Balakrishnan, "Analysis of Aircraft Fuel Burn and Emissions in the Landing and Take Off Cycle Using Operational Data," 6th International Conference on Research in Air Transportation (ICRAT 2014), May 10, 2014, http://www.mit.edu/~hamsa/pubs/ICRAT_2014_YSC_HB_final.pdf.

19. Slow Dinosaur Apocalypse

Crosta, G. B., P. Frattini, E. Valbuzzi, and F. V. De Blasio, "Introducing a New Inventory of Large Martian Landslides," *Earth and Space Science* 5, no. 4 (March 1, 2018): 89–119, https://agupubs.onlinelibrary.wiley.com/doi/full/10.1002/2017EA000324.

DePalma, Robert A., Jan Smit, David A. Burnham, Klaudia Kuiper, Phillip L. Manning, Anton Oleinik, Peter Larson, Florentin J. Maurrasse, Johan Vellekoop, Mark A. Richards, Loren Gurche, and Walter Alvarez, "A Seismically Induced Onshore Surge Deposit at the KPg Boundary, North Dakota," *PNAS* 116, no. 7 (April 1, 2019): 8190–8199, https://doi.org/10.1073/pnas.1817407116.

Korycansky, D. G., and Patrick J. Lynett, "Run-up from Impact Tsunami," *Geophysical Journal International* 170, no. 3 (September 1, 2007): 1076–1088, https://doi.org/10.1111/j.1365-246X.2007.03531.x.

Massel, Stanisław R., "Tsunami in Coastal Zone Due to Meteorite Impact," *Coastal Engineering* 66, (2012): 40–49, https://doi.org/10.1016/j.coastaleng.2012.03.013.

Schulte, Peter, Jan Smit, Alexander Deutsch, Tobias Salge, Andrea Friese, and Kilian Beichel, "Tsunami Backwash Deposits with Chicxulub Impact Ejecta and Dinosaur Remains from the Cretaceous–Palaeogene Boundary in the La Popa Basin, Mexico," *Sedimentology* 59, no. 3 (April 1, 2012): 737–765, doi:10.1111/j.1365-3091.2011.01274.x.

Su, Xing, Wanhong Wei, Weilin Ye, Xingmin Meng, and Weijiang Wu, "Predicting Landslide Sliding Distance Based on Energy Dissipation and Mass Point Kinematics," *Natural Hazards* 96 (2019): 1367–1385, https://doi.org/10.1007/s11069-019-03618-z.

Wünnemann, K., and R. Weiss, "The Meteorite Impact-Induced Tsunami Hazard," *The Royal Society* 373, no. 2053 (October 28, 2015), https://doi.org/10.1098/rsta.2014.0381.

23. $2 Undecillion Lawsuit

Boston Consulting Group: Press Releases, "Despite COVID-19, Global Financial Wealth Soared to Record High of $250 Trillion in 2020," June 10, 2021, https://www.bcg.com/press/10june2021-despite-covid-19-global-financial-wealth-soared-record-high-250-trillion-2020.

24. Star Ownership

White, Reid, "Plugging the Leaks in Outer Space Criminal Jurisdiction: Advocation for the Creation of a Universal Outer Space Criminal Statute," *Emory International Law Review* 35, no. 2 (2021), https://scholarlycommons.law.emory.edu/eilr/vol35/iss2/5/.

25. Tire Rubber

Halle, Louise L., Annemette Palmqvist, Kristoffer Kampmann, and Farhan R. Khana, "Ecotoxicology of Micronized Tire Rubber: Past, Present and Future Considerations," *Science of the Total Environment* 706, no. 1, (March 2020), https://doi.org/10.1016/j.scitotenv.2019.135694.

Parker-Jurd, Florence N. F., Imogen E. Napper, Geoffrey D. Abbott, Simon Hann, Richard C. Thompson, "Quantifying the Release of Tyre Wear Particles to the Marine Environment Via Multiple Pathways," *Marine Pollution Bulletin* 172 (November 2021), https://www.sciencedirect.com/science/article/abs/pii/S0025326X21009310.

Sieber, Ramona, Delphine Kawecki, and Bernd Nowack, "Dynamic Probabilistic Material Flow Analysis of Rubber Release from Tires into the Environment," *Environmental Pollution* 258 (March 2020), https://www.sciencedirect.com/science/article/abs/pii/S0269749119333998.

Tian, Zhenyu, Haoqi Zhao, Katherine T. Peter, Melissa Gonzalez, Jill Wetzel, Christopher Wu, Ximin Hu, Jasmine Prat, Emma Mudrock, Rachel Hettinger, Allan E. Cortina, Rajshree Ghosh Biswas, Flávio Vinicius Crizóstomo Kock, Ronald Soong, Amy Jenne, Bowen Du, Fan Hou, Huan He, Rachel Lundeen, Alicia Gilbreath, Rebecca Sutton, Nathaniel L. Scholz, Jay W. Davis, Michael C. Dodd, Andre Simpson, Jenifer K. McIntyre, and Edward P. Kolodziej, "A Ubiquitous Tire Rubber–Derived Chemical Induces Acute Mortality in Coho Salmon," *Science* 371, no. 6525 (January 8, 2021): 185–189, https://www.science.org/doi/abs/10.1126/science.abd6951.

26. Plastic Dinosaurs

Fuel Chemistry Division, "Petroleum," https://personal.ems.psu.edu/~pisupati/ACSOutreach/Petroleum_2.html.

Goñi, Miguel A., Kathleen C. Ruttenberg, and Timothy I. Eglinton, "Sources and Contribution of Terrigenous Organic Carbon to Surface Sediments in the Gulf of Mexico," *Nature* 389 (1997): 275–278, https://www.whoi.edu/cms/files/goni_et_al_Nature_1997_35805.pdf.

Libes, Susan, "The Origin of Petroleum in the Marine Environment," chap. 26 in *Introduction to Marine Biogeochemistry* (Cambridge, MA: Elsevier, 2009), https://booksite.elsevier.com/9780120885305/casestudies/01-Ch26-P088530web.pdf.

Powell, T. G., "Developments in Concepts of Hydrocarbon Generation from Terrestrial Organic Matter," 1989, https://archives.datapages.com/data/circ_pac/0011/0807_f.htm.

State of Louisiana: Department of Natural Resources, "Where Does Petroleum Come From? Why Is It Normally Found in Huge Pools Under Ground? Was It Formed in a Big Pool Where We Find It, or Did It Gather There Due to Outside Natural Forces?," http://www.dnr.louisiana.gov/assets/TAD/education/BGBB/3/origin.html.

University of South Carolina, "School of the Earth, Ocean, and Environment," https://sc.edu/study/colleges_schools/artsandsciences/earth_ocean_and_environment/index.php.

27. Suction Aquarium

Bailey, Helen, and David H. Secor, "Coastal Evacuations by Fish During Extreme Weather Events," *Sci Rep* 6, no. 30280 (2016), https://doi.org/10.1038/srep30280.

Brown, Frank A., Jr., "Responses of the Swimbladder of the Guppy, Lebistes reticulatus, to Sudden Pressure Decreases," *The Biological Bulletin* 76, no. 1 (1939): 48–58, https://www.jstor.org/stable/1537634.

Heupel, M. R., C. A. Simpfendorfer, and R. E. Hueter, "Running Before the Storm: Blacktip Sharks Respond to Falling Barometric Pressure Associated with Tropical Storm Gabrielle," *Journal of Fish Biology* 63 (2003): 1357–1363, https://onlinelibrary.wiley.com/doi/abs/10.1046/j.1095-8649.2003.00250.x.

Hogan, Joe, "The Effects of High Vacuum on Fish," *Transactions of the American Fisheries Society* 70, no. 1 (1941): 469–474, https://afspubs.onlinelibrary.wiley.com/doi/abs/10.1577/1548-8659%281940%2970%5B469%3ATEOHVO%5D2.0.CO%3B2.

Holbrook, R. I., and T. B. de Perera, "Fish Navigation in the Vertical Dimension: Can Fish Use Hydrostatic Pressure to Determine Depth?," *Fish and Fisheries* 12 (2011): 370–379, https://onlinelibrary.wiley.com/doi/10.1111/j.1467-2979.2010.00399.x.

Sullivan, Dan M., Robert W. Smith, E. J. Kemnitz, Kevin Barton, Robert M. Graham, Raymond A. Guenther, and Larry Webber, "What Is Wrong with Water Barometers?," *The Physics Teacher* 48, no. 3 (2010): 191–193, https://aapt.scitation.org/doi/10.1119/1.3317456.

28. Earth Eye

Mishima, S., A. Gasset, S. D. Klyce, and J. L. Baum, "Determination of Tear Volume and Tear Flow," *Invest. Ophthalmol. Vis. Sci.* 5, no. 3 (1966): 264–276, https://iovs.arvojournals.org/article.aspx?articleid=2203634.

Steinbring, Eric, "Limits to Seeing High-Redshift Galaxies Due to Planck-Scale-Induced Blurring," *Proceedings of the International Astronomical Union* 11, no. S319, 54–54, 2015, doi:10.1017/S1743921315009850.

29. Build Rome in a Day

The Civic Federation, "Estimated Full Value of Real Estate in Cook County Saw Six Straight Years of Growth Between 2012–2018," October 30, 2020, https://www.civicfed.org/civic-federation/blog/estimated-full-value-real-estate-cook-county-saw-six-straight-years-growth.

U.S. Bureau of Economic Analysis, "Gross Domestic Product: All Industries in Cook County, IL [GDPALL17031]," retrieved from FRED, Federal Reserve Bank of St. Louis, November 20, 2021, https://fred.stlouisfed.org/series/GDPALL17031.

30. Mariana Trench Tube

Stommel, Henry, Arnold B. Arons, and Duncan Blanchard, "An Oceanographical Curiosity: The Perpetual Salt Fountain," *Deep Sea Research* 3, no. 2 (1956): 152–153, https://www.sciencedirect.com/science/article/pii/0146631356900958.

31. Expensive Shoebox

"What is the volume of a kilogram of cocaine?," The Straight Dope Message Board, https://boards.straightdope.com/t/what-is-the-volume-of-a-kilogram-of-cocaine/286573.

32. MRI Compass

NOAA, "Maps of Magnetic Elements from the WMM2020," https://www.ngdc.noaa.gov/geomag/WMM/image.shtml.

Tremblay, Charles, Sylvain Martel, binjamin conan, Dumitru Loghin, and alexandre bigot, "Fringe Field Navigation for Catheterization," *IFMBE Proceedings* 45 (2014), https://www.researchgate.net/publication/270759488_Fringe_Field_Navigation_for_Catheterization.

33. Ancestor Fraction

Kaneda, Toshiko, and Carl Haub, "How Many People Have Ever Lived on Earth?," *Population Reference Bureau*, May 18, 2021, https://www.prb.org/articles/how-many-people-have-ever-lived-on-earth/.

Rohde, Douglas L. T., Steve Olson, and Joseph T. Chang, "Modelling the Recent Common Ancestry of All Living Humans," *Nature* 431 (2004): 562–566, https://doi.org/10.1038/nature02842.

Roser, Max, "Mortality in the Past—Around Half Died As Children," *Our World in Data*, June 11, 2019, https://ourworldindata.org/child-mortality-in-the-past.

34. Bird Car

Mosher, James A., and Paul F. Matray, "Size Dimorphism: A Factor in Energy Savings for Broad-Winged Hawks," *The Auk* 91, no. 2 (April 1974): 325–341, https://www.jstor.org/stable/4084511.

Pennycuick, C. J., Holliday H. Obrecht III, and Mark R. Fuller, "Empirical Estimates of Body Drag of Large Waterfowl and Raptors," *J Exp Biol* 135, no. 1 (March 1988): 253–264, https://journals.biologists.com/jeb/article/135/1/253/5435/Empirical-Estimates-of-Body-Drag-of-Large.

35. No-Rules NASCAR

Kumar, Vasantha K., and William T. Norfleet, "Issues on Human Acceleration Tolerance After Long-Duration Space Flights," NASA Technical Memorandum 104753, October 1, 1992, https://ntrs.nasa.gov/citations/19930020462.

National Aeronautics and Space Administration, "Astronautics and its Applications," Environment of Manned Systems: Internal Environment of Manned Space Vehicles, 105–126, https://history.nasa.gov/conghand/mannedev.htm.

Spark, Nick T., "46.2 Gs!!!: The Story of John Paul Stapp, 'The Fastest Man on Earth,'" *Wings/Airpower Magazine*, http://www.ejectionsite.com/stapp.htm.

36. Vacuum Tube Smartphone

Shilov, Anton, "Apple's A14 SoC Under the Microscope: Die Size & Transistor Density Revealed," Tom's Hardware, October 29, 2020, https://www.tomshardware.com/news/apple-a14-bionic-revealed.

Sylvania, "Engineering Data Service," http://www.nj7p.org/Tubes/PDFs/Frank/137-Sylvania/7AK7.pdf.

War Department: Bureau of Public Relations, "Physical Aspects, Operation of ENIAC are Described," February 16, 1946, https://americanhistory.si.edu/comphist/pr4.pdf.

37. Laser Umbrella

Hautière, Nicholas, Eric Dumont, Roland Brémond, and Vincent Ledoux, "Review of the Mechanisms of Visibility Reduction by Rain and Wet Road," ISAL Conference, 2009, https://www.researchgate.net/publication/258316669_Review_of_the_Mechanisms_of_Visibility_Reduction_by_Rain_and_Wet_Road.

Pendleton, J. D., "Water Droplets Irradiated by a Pulsed CO_2 Laser: Comparison of Computed Temperature Contours with Explosive Vaporization Patterns," *Applied Optics* 24, no. 11 (1985): 1631–1637, https://www.osapublishing.org/ao/abstract.cfm?uri=ao-24-11-1631.

Sageev, Gideon, and John H. Seinfeld, "Laser Heating of an Aqueous Aerosol Particle," *Applied Optics* 23, no. 23 (December 1, 1984), http://authors.library.caltech.edu/10136/1/SAGa084.pdf.

Takamizawa, Atsushi, Shinji Kajimoto, Jonathan Hobley, Koji Hatanaka, Koji Ohtab, and Hiroshi Fukumura, "Explosive Boiling of Water After Pulsed IR Laser Heating," *Physical Chemistry Chemical Physics* 5 (2003), https://pubs.rsc.org/en/content/articlelanding/2003/CP/b210609d.

40. Lava Lamp

UNEP Chemicals Branch, "The Global Atmospheric Mercury Assessment: Sources, Emissions and Transport," UNEP-Chemicals, Geneva, 2008, https://wedocs.unep.org/bitstream/handle/20.500.11822/13769/UNEP_GlobalAtmosphericMercuryAssessment_May2009.pdf?sequence=1&isAllowed=y.

41. Sisyphean Refrigerators

Thurber, Caitlin, Lara R. Dugas, Cara Ocobock, Bryce Carlson, John R. Speakman, and Herman Pontzer, "Extreme Events Reveal an Alimentary Limit on Sustained Maximal Human Energy Expenditure," *Science Advances* 5, no. 6, https://www.science.org/doi/10.1126/sciadv.aaw0341.

42. Blood Alcohol

Thank you to Conor Braman, among others, for correcting a missing zero in the original version of this chapter's calculations.

Brady, Ruth, Sara Suksiri, Stella Tan, John Dodds, and David Aine, "Current Health and Environmental Status of the Maasai People in Sub-Saharan Africa," *Cal Poly Student Research: Honors Journal 2008*, 17–32, https://digitalcommons.calpoly.edu/cgi/viewcontent.cgi?referer=&httpsredir=1&article=1005&context=honors.

United States Air Force Medical Service, "Alcohol Brief Counseling: Alcohol Education Module," Air Force Alcohol and Drug Abuse Prevention and Treatment Tier II, October 2007, https://www.minot.af.mil/Portals/51/documents/resiliency/AFD-111004-028.pdf?ver=2016-06-10-110043-200.

44. Spiders vs. the Sun

Greene, Albert, Jonathan A. Coddington, Nancy L. Breisch, Dana M. De Roche, and Benedict B. Pagac Jr., "An Immense Concentration of Orb-Weaving Spiders with Communal Webbing in a Man-Made Structural Habitat (Arachnida: Araneae: Tetragnathidae, Araneidae)," *American Entomologist: Fall 2010*, 146–156, https://www.entsoc.org/PDF/2010/Orb-weaving-spiders.pdf.

Höfer, Hubert and Ricardo Ott, "Estimating Biomass of Neotropical Spiders and Other Arachnids (Araneae, Opiliones, Pseudoscorpiones, Ricinulei) by Mass-Length Regressions," *The Journal of Arachnology* 37, no. 2 (2009): 160–169, https://doi.org/10.1636/T08-21.1.

Newman, Jonathan A., and Mark A. Elgar, "Sexual Cannibalism in Orb-Weaving Spiders: An Economic Model," *The American Naturalist* 138, no. 6 (1991): 1372–1395, https://www.jstor.org/stable/2462552.

Topping, Chris J., and Gabor L. Lovei, "Spider Density and Diversity in Relation to Disturbance in Agroecosystems in New Zealand, with a Comparison to England," *New Zealand Journal of Ecology* 21, no. 2 (1997): 121–128, https://newzealandecology.org/nzje/2020.

Wilder, Shawn M. and Ann L. Rypstra, "Trade-off Between Pre- and Postcopulatory Sexual Cannibalism in a Wolf Spider (Araneae, Lycosidae)," *Behavioral Ecology and Sociobiology* 66 (2012): 217–222, https://link.springer.com/article/10.1007/s00265-011-1269-0.

45. Inhale a Person

Clark, R. P., and S. G. Shirley, "Identification of Skin in Airborne Particulate Matter," *Nature* 246 (1973): 39–40, https://www.nature.com/articles/246039a0.

Morawska, Lidia and Tunga Salthammer, eds., *Indoor Environment: Airborne Particles and Settled Dust* (Hoboken, NJ: Wiley, 2003).

Weschler, Charles J., Sarka Langer, Andreas Fischer, Gabriel Bekö, Jørn Toftum, and Geo Clausen, "Squalene and Cholesterol in Dust from Danish Homes and Daycare Centers," *Environ. Sci. Technol.* 45, no. 9 (2011): 3872–3879, https://pubs.acs.org/doi/10.1021/es103894r.

46. Candy Crush Lightning

Xie, Yujun, and Zhen Li, "Triboluminescence: Recalling Interest and New Aspects," *Chem* 4, no. 5 (May 10, 2018), https://doi.org/10.1016/j.chempr.2018.01.001.

S Short Answers #4

Ratnayake, Wajira S., and David S. Jackson, "Gelatinization and Solubility of Corn Starch During Heating in Excess Water: New Insights," *Journal of Agricultural and Food Chemistry* 54, no. 10 (2006): 3712–3716, https://pubs.acs.org/doi/10.1021/jf0529114.

Wertheim, Heiman F. L., Thai Q. Nguyen, Kieu Anh T. Nguyen, Menno D. de Jong, Walter R. J. Taylor, Tan V. Le, Ha H. Nguyen, Hanh T. H. Nguyen, Jeremy Farrar, Peter Horby, and Hien D. Nguyen, "Furious Rabies After an Atypical Exposure," *PLoS Med.* 6, no. 3 (2009): e1000044, https://doi.org/10.1371/journal.pmed.1000044.

48. Proton Earth, Electron Moon

Carroll, Sean, "The Universe Is Not a Black Hole," 2010, http://www.preposterousuniverse.com/blog/2010/04/28/the-universe-is-not-a-black-hole/.

Garon, Todd S., and Nelia Mann, "Re-examining the Value of Old Quantization and the Bohr Atom Approach," *American Journal of Physics* 81, no. 2, (2013): 92, https://aapt.scitation.org/doi/10.1119/1.4769785.

50. Japan Runs an Errand

Lindsey, Rebecca, "Climate Change: Global Sea Level," Climate.gov, August 14, 2020, https://www.climate.gov/news-features/understanding-climate/climate-change-global-sea-level.

Gamo, T., N. Nakayama, N. Takahata, Y. Sano, J. Zhang, E. Yamazaki, S. Taniyasu, and N. Yamashita, "Revealed by Time-Series Observations over the Last 30 Years," 2014, https://www.semanticscholar.org/paper/Revealed-by-Time-Series-Observations-over-the-Last-Gamo-Nakayama/57bd09d9b01e7735cd593b5a2147a9c64bbd5b7e?p2df.

Ward, Steven N., and Erik Asphaug, "Impact Tsunami-Eltanin," *Deep-Sea Research II* 49 (2002): 1073–1079, https://websites.pmc.ucsc.edu/~ward/papers/final_eltanin.pdf.

51. Fire from Moonlight

Plait, Phil, "BAFact Math: The Sun Is 400,000 Times Brighter than the Full Moon," *Discover Magazine: Bad Astronomy*, August 27, 2012, https://www.discovermagazine.com/the-sciences/bafact-math-the-sun-is-400-000-times-brighter-than-the-full-moon.

52. Read All the Laws

FindLaw, "California Code, Food and Agricultural Code (Formerly Agricultural Code)—FAC § 27637," https://codes.findlaw.com/ca/food-and-agricultural-code-formerly-agricultural-code/fac-sect-27637.html.

Fish, Eric S., "Judicial Amendment of Statutes," 84 *George Washington Law Review* 563 (2016), https://papers.ssrn.com/sol3/papers.cfm?abstract_id=2656665.

GovInfo, "F Code of Federal Regulations (Annual Edition)," https://www.govinfo.gov/app/collection/cfr.

Legal Information Institute, "Primary Authority," Cornell Law, https://www.law.cornell.edu/wex/primary_authority.

U.S. Department of State, "Treaties in Force," Office of Treaty Affairs, https://www.state.gov/treaties-in-force/.

Zittrain, Jonathan, "The Supreme Court and Zombie Laws," July 2, 2018, https://medium.com/@zittrain/the-supreme-court-and-zombie-laws-2087d7bb9a75.

53. Saliva Pool

Fédération Internationale de Natation, "FR 2: Swimming Pools," https://web.archive.org/web/20160902023159/http://www.fina.org/content/fr-2-swimming-pools.

Watanabe, S., M. Ohnishi, K. Imai, E. Kawano, and S. Igarashi, "Estimation of the Total Saliva Volume Produced Per Day in Five-Year-Old Children," *Arch Oral Biol.* 40, no. 8, 781–782, https://www.sciencedirect.com/science/article/abs/pii/000399699500026L?via%3Dihub.

55. Niagara Straw

Cashco, "Fluid Flow Basics of Throttling Valves," 17, https://www.controlglobal.com/assets/Media/MediaManager/RefBook_Cashco_Fluid.pdf.

New York Power Authority, "Niagara River Water Level and Flow Fluctuations Study Final Report," *Niagara Power Project FERC No. 2216*, August 2005, https://web.archive.org/web/20160229090220/http://niagara.nypa.gov/ALP%20working%20documents/finalreports/html/IS23WL.htm.

56. Walking Backward in Time

Blum, M. D., M. J. Guccione, D. A. Wysocki, P. C. Robnett, E. M. Rutledge, "Late Pleistocene Evolution of the Lower Mississippi River Valley, Southern Missouri to Arkansas," *GSA Bulletin* 112, no. 2 (February 2000): 221–235, https://pubs.geoscienceworld.org/gsa/gsabulletin/article-abstract/112/2/221/183594/Late-Pleistocene-evolution-of-the-lower?redirectedFrom=fulltext.

Braun, Duane D., "The Glaciation of Pennsylvania, USA," *Developments in Quaternary Sciences* 15 (2011): 521–529, https://www.sciencedirect.com/science/article/abs/pii/B9780444534477000404.

Bryant, Jr., Vaughn M., "Paleoenvironments," Handbook of Texas Online, 1995, https://www.tshaonline.org/handbook/entries/paleoenvironments.

Carson, Eric C., J. Elmo Rawling III, John W. Attig, and Benjamin R. Bates, "Late Cenozoic Evolution of the Upper Mississippi River, Stream Piracy, and Reorganization of North American Mid-Continent Drainage Systems," *GSA Today* 28, no. 7 (July 2018): 4–11, https://www.geosociety.org/gsatoday/science/G355A/abstract.htm.

Fildani, Andrea, Angela M. Hessler, Cody C. Mason, Matthew P. McKay, and Daniel F. Stockli, "Late Pleistocene Glacial Transitions in North America Altered Major River Drainages, as Revealed by Deep-Sea Sediment," *Scientific Reports* 8 (2018), https://www.nature.com/articles/s41598-018-32268-7.

"Interglacials of the Last 800,000 Years," *Reviews of Geophysics* 54, no. 1 (2015): 162–219, https://agupubs.onlinelibrary.wiley.com/doi/10.1002/2015RG000482.

Knox, James C., "Late Quaternary Upper Mississippi River Alluvial Episodesa Their Significance to the Lower Mississippi River System," *Engineering Geology* 45, no. 1–4 (December 1996): 263–285, https://www.sciencedirect.com/science/article/abs/pii/S0013795296000178?via%3Dihub.

Millar, Susan W. S., "Identification of Mapped Ice-Margin Positions in Western New York from Digital Terrain-Analysis and Soil Databases," *Physical Geography* 25, no. 4 (2004): 347–359, https://www.tandfonline.com/doi/abs/10.2747/0272-3646.25.4.347.

Sheldon, Robert A., *Roadside Geology of Texas* (Missoula, MT: Mountain Press Publishing Company, 1991).

57. Ammonia Tube

Padappayil, Rana Prathap, and Judith Borger, "Ammonia Toxicity," StatPearls Publishing LLC, https://www.ncbi.nlm.nih.gov/books/NBK546677/.

Short Answers #5

Olive Garden, "Nutrition Information," https://media.olivegarden.com/en_us/pdf/olive_garden_nutrition.pdf.

Sagar, Stephen M., Robert J. Thomas, L. T. Loverock, and Margaret F. Spittle, "Olfactory Sensations Produced by High-Energy Photon Irradiation of the Olfactory Receptor Mucosa in Humans," *International Journal of Radiation Oncology, Biology, Physics* 20, no. 4 (April 1991): 771–776, https://www.sciencedirect.com/science/article/abs/pii/036030169190021U.

59. Global Snow

Buckler, J. M., "Variations in Height Throughout the Day," *Archives of Disease in Childhood* 53, no. 9 (1989): 762, http://dx.doi.org/10.1136/adc.53.9.762.

National Oceanic and Atmospheric Administration, "Welcome to: Cooperative Weather Observer: Snow Measurement Training," National Weather Service, https://web.archive.org/web/20150221171450/http://www.srh.noaa.gov/images/mrx/coop/SnowMeasurementTraining.pdf.

Roylance, Frank D., "A Likely Record, but Experts Will Get Back to Us," *Baltimore Sun*, https://web.archive.org/web/20140716134151/http://articles.baltimoresun.com/2010-02-07/news/bal-md.storm07feb07_1_baltimore-washington-forecast-office-snow-depth-biggest-storm.

61. Into the Sun

IEEE, org, "IEEE 1584-2018, IEEE Guide for Performing Arc-Flash Hazard Calculations," https://www.techstreet.com/ieee/standards/ieee-1584-2018?gateway_code=ieee&vendor_id=5802&product_id=1985891.

62. Sunscreen

Food and Drug Administration, "Sunscreen Drug Products," https://www.regulations.gov/docket/FDA-1978-N-0018.

63. Walking on the Sun

Blouin, S., P. Dufour, C. Thibeault, and N. F. Allard, "A New Generation of Cool White Dwarf Atmosphere Models. IV. Revisiting the Spectral Evolution of Cool White Dwarfs," *The Astrophysical Journal* 878, no. 1 (2019), https://iopscience.iop.org/article/10.3847/1538-4357/ab1f82.

Chen, Eugene Y., and Brad M. S. Hansen, "Cooling Curves and Chemical Evolution Curves of Convective Mixing White Dwarf Stars," *Monthly Notices of the Royal Astronomical Society* 413, no. 4 (June 2011): 2827–2837, https://academic.oup.com/mnras/article/413/4/2827/965051.

Koberlein, Brian, "Frozen Star," March 2, 2014, https://briankoberlein.com/blog/frozen-star/.

Renedo, I., L. G. Althaus, M. M. Miller Bertolami, A. D. Romero, A. H. Córsico, R. D. Rohrmann, and E. García-Berro, "New Cooling Sequences for Old White Dwarfs," *The Astrophysics Journal* 717, no. 1 (2010), https://iopscience.iop.org/article/10.1088/0004-637X/717/1/183.

Salaris, M., L. G. Althaus, and E. García-Berro, "Comparison of Theoretical White Dwarf Cooling Timescales," *Astronomy & Astrophysics* 555 (July 2013), https://www.aanda.org/articles/aa/full_html/2013/07/aa20622-12/aa20622-12.html.

Srinivasan, Ganesan, *Life and Death of the Stars*, Undergraduate Lecture Notes in Physics, 2014, https://link.springer.com/book/10.1007/978-3-642-45384-7.

Veras, Dimitri, and Kosuke Kurosawa, "Generating Metal-Polluting Debris in White Dwarf Planetary Systems from Small-Impact Crater Ejecta," *Monthly Notices of the Royal Astronomical Society* 494, no. 1 (May 2020): 442–457, https://academic.oup.com/mnras/article-abstract/494/1/442/5788436?redirectedFrom=fulltext.

Wilson, R. Mark, "White Dwarfs Crystallize as They Cool," *Physics Today* 72, no. 3 (2019): 14, https://physicstoday.scitation.org/doi/10.1063/PT.3.4156.

64. Lemon Drops and Gumdrops

Goldblatt, C., T. Robinson, and D. Crisp, "Low Simulated Radiation Limit for Runaway Greenhouse Climates," *Nature Geoscience* 6 (2013): 661–667, https://www.semanticscholar.org/paper/Low-simulated-radiation-limit-for-runaway-climates-Goldblatt-Robinson/4be39d2e4114f1347569d81029f59005e141befe.

Gunina, Anna, and Yakov Kuzyakov, "Sugars in Soil and Sweets for Microorganisms: Review of Origin, Content, Composition and Fate," *Soil Biology and Biochemistry* 90 (2015): 87–100, https://www.sciencedirect.com/science/article/abs/pii/S0038071715002631.

Heymsfield, Andrew J., Ian M. Giammanco, and Robert Wright, "Terminal Velocities and Kinetic Energies of Natural Hailstones," *Geophysical Research Letters* 41, no. 23 (November 25, 2014): 8666–8672, https://agupubs.onlinelibrary.wiley.com/doi/full/10.1002/2014GL062324.

Myhre, G., D. Shindell, F.-M. Bréon, W. Collins, J. Fuglestvedt, J. Huang, D. Koch, J.-F. Lamarque, D. Lee, B. Mendoza, T. Nakajima, A. Robock, G. Stephens, T. Takemura, and H. Zhang, "Anthropogenic and Natural Radiative Forcing," *Climate Change 2013: The Physical Science Basis*, https://www.ipcc.ch/site/assets/uploads/2018/02/WG1AR5_Chapter08_FINAL.pdf.

Index

Italicized page numbers indicate material in illustrations.

A

acceleration, *181*
of helicopter rotors, 6–7, 9–10
human tolerance for, 94, 180–83, *182*
accelerators, 184, 237–38
agriculture, 331
air. *See also* atmosphere
air pressure, 145–48, 160
in clouds, 195–96
cold, heat, and, 13–14, 31, *31*, 44–45, *45*
sword of, 31, *31*
Air Force One (aircraft), 51
air hockey table, house floor as, 230
airplanes, 83–87
airspace, extending, 124–27, *125*
alcohol, in blood, 210–11, 211*n*, 213
ammonia, 284–86
ancestors, 172–75, *173*, *174*, *175*
ants, 135–36, 262
anvil, dropping from space, 233
aquariums, 143–48
astronauts, 319
atmosphere, 145
iron in, *18*, 18–20
light in, *44*, 44–46, 44*n*
atoms, 36, 142, 236–37, 236*n*
Au Bon Pain, 120–22
Austin, Texas, 277–80
Australia, 26, 28–29, 124–25, *125*

B

Back River Wastewater Treatment Plant, spiders in, 219, 219*n*
baking, with lasers, 228
ball, of sand, 99–100
balloon, hot-air, 56, *56*
balloons, 88
camera on, 32, *32*
on Venus, *233*
snakes swallowing, 33, *33*
bananas, 52–54, *54*
barometers, 147
basketball, spinning, 214
bees, 51
birds, *29*. *See also* pigeons
car catching, 176–79, *177*, *178*
flight of, 26–29, 27*n*, 177–78, *178*, 178*n*, 231, *231*
black holes, *2*, 2–5, 125
Cygnus X-1, 126, 126*n*
electron Moon and, 239–40, 239*n*
blades, helicopter. *See* helicopters
Blechner, A. J., 259
blood
alcohol in, 210–11, 211*n*, 213
ants in, 262
drinking, eating, 210–13, 211*n*
liquid uranium in, 30
body
bones in, breaking, removing, 92, 94
water in, 31, *32*, 134
weight of, Earth and, 59–65, 108, 108*n*
boiling, *16*, 16–17, 273
bones, 92, 94
books, 47–50
Boötes, 152–53
breadsticks, chewing, 299–300, *300*
breathing, 19
buildings, size of, 111, *114*. *See also* Q1 skyscraper; Rome, building
elevators and, 112–13, 119
mega-skyscrapers, *115*, 115–18, *116*, *117*, *118*
bullets, 55–58, 55*n*, 58*n*, 138

C

California Food and Agricultural Code § 27637, 258, 260
calories, 37–39, 65
camera, 32, *32*
The Canterbury Tales (Chaucer), 47–48
car, catching birds in, 176–79, *177*, *178*
carbon dioxide, 132*n*, 332–34, 333*n*
car racing, *180*, 180–84
case law, 257, 257*n*
catapult, 83–87
cavitation bubbles, 273, 273*n*
centrifugal force, 8, 108, 108*n*
Ceres, 103
cerium, 103
Chan, Daniel W. M., 154–55
Chaucer, Geoffrey. *See The Canterbury Tales*
chewing, 299–300, *300*
Chicxulub impact, 98, 100–102, *101*, 215
chlorine gas, 284–85
churches, 52–54
climate change, 20, *21*, *208*, 209, 228, 280. *See also* global warming
clouds, eating, 195–98
coal, 131, 133
cocaine, 166–67, 166*n*
cold, 12, *12*, *13*, 206
air and, 13–14, 31, *31*, 45, *45*
in deep ocean, 159–60
in space, 93–94

cold *(cont.)*
ultracold materials, 11, 14–15, *15*
compasses, 168–71
computers, early, 186–87, *187*
conservation of étendue, 253, *253*, 253n
Constitution, US, 255–56
construction projects, duration of, 154–55
The Core (film), 227
corona, of Sun, 316n
cotton candy, 197
crash rates, 22–23
currents, ocean, 247, *247*
Cygnus X-1, 126, 126n

D

day, length of, 215–16
Daytona International Speedway, *180*, 180–81, 183–84
dead skin, 220–22, 220n
Death in Yellowstone (Whittlesey), 40–41
Death on the High Seas Act (DOHSA), 127
deep ocean, 159–61
defibrillator, 35
density
of cocaine, 166, 166n
of Earth, 63
of Jupiter, 71
of water, 71, 197, 301, 301n, 332, 332n
Department of Energy, 258, 258n
diamonds, 166, 184
diffraction, 150–51
dinosaurs, 37n, 133n
metabolism of, 37–39, 38n
meteors and extinction of, 98, 102, 138n, 215
in oil, dead, 131–34
diving bell spider, 218n
dogs, 308–13, 309n
DOHSA. *See* Death on the High Seas Act
double vision, 241–43
driven oscillation, 79
driving, 23n, 132
acceleration in, 180–82, *181*
car racing, *180*, 180–84
catching birds while, 176–79, *177*, *178*
to edge, of universe, 22–25
mailbox punched during, 185
drugs, price of, 166–67
Dunbar, Robin, 24
Dungeons & Dragons, 140
Durban, South Africa, 81
dust, dead skin in, 220–22, 220n
Dyson sphere, 61–62, 312

E

Earth, 2, *3*, *124*
airspace of, extending, 124–27, *125*
body weight and, 59–65, 108, 108n
crust, mantle of, 61–63, 109, 121, 121n, 191, 227
elements on, 121, 121n
as eye, massive, 149–53
gravity of, 3, 59, 63–65, 287, 289–90, 289n, 326–27, *327*
magnetic fields of, 228
from Moon, pole to, 287–97, *289*, *293*
as perfect sphere, 136
of protons, 236–37, 239n
rotation of, 108–10, 108n
slicing in half, 229
snow covering, 305–7
spin of, 214–16, *215*, 287, 293, 293n
sugar covering, 332
Sun consuming, 323n
surface of, removing, 61–63
without water, 34, *34*
economic value
insects creating, 96
total, 120–22, 121n
edge, of observable universe, 22–25
eggs, 250, 258, 300–301, 300n
electricity, 207, 234–35, *235*
electron beams, 17, *17*
electrons, Moon made of, 236–40, 239n
elements, on Earth, 121, 121n
elephant, 37–38, 37n
elevators, 112–13, 119, 137
engine oil, 23, *23*, 23n
English, books in, 47–48, 50, 1–334
ENIAC, 187, *187*
epinephrine, 185
equator, 293, 293n
evaporation, of iron, 16–21
event horizon, 2, *2*, 4
extinction, 98, 102, 138–39, 138n, 215
eye, *152*, 243n, 278. *See also* visibility distance; visible objects, in space
Earth as massive, 149–53
eyeball, pointing eyeball at, 241–43

F

family trees, 172–74, *173*, *174*
Fata Morgana, 46
Fermi estimation, 66, 66n, *67*, 68–69
fire, 141–42, 141n, 249–51, 254
fish, 145–47
Fletcher, Horace, 300, *300*
flicker fusion threshold, 278
flight
bird, 26–29, 27n, 177–78, *178*, 178n, 231, *231*
insect, 90, *90*
pigeon, 26–29, 27n, 231, *231*
flow rate, 273–75, 273n
fluids, pumping, 273, 273n
47 Ursae Majoris, 126–27
fossil fuels, 59–60, 83, 121n. *See also* fuel
burning, 132, 132n
dinosaurs in, 131–34
seizing, 59
free-hanging length, *270*, 270–71
fringe fields, magnetic, 168–69
fuel
airplanes using, 83–84
for driving, 23, 23n, 132
fused quartz glass, 202–3

G

galactic core, 125–26, *126*
gas
  chlorine, 284–85
  in water, 148
gasoline, 23, *23*, 23*n*
general relativity, 4–5, 237–39
geysers, 40–42, *41*, *42*
g-forces, 94, 180–83
glacial periods, 282
glass beaches, 247, 247*n*
global warming, 228, 332–34
  iron and, 20, *21*
  refrigerators and, 207, *208*, 209
gold, 165–66, 204*n*
goliath bird spider, 218, 218*n*
grains, of sand, 74–77
grapes, 36
gravity
  birds and, 231
  black holes and, 239–40
  body weight and, 59, 63
  of Earth, 3, 59, 63–65, 287, 289–90, 289*n*, 326–27, *327*
  electrons and, 237, 239–40
  in free-hanging length, 270, *270*
  Japan, disappearance of, and, 245, 245*n*
  Jupiter and, 71–73, *72*
  on Moon, 287–89
  snowballs and, 269–70
  spiders, gravitational pull of, 217–19
  Sun, gravitational pull of, 217–19, 323
  on white dwarf, 326–27, *327*
GRB 080319b, 152–53
Great Lakes–St. Lawrence River Adaptive Management Committee, 272, 272*n*
grenade, 93
gumdrops, rain of lemon drops and, 329–44
guns, 91

H

hamburgers, 39
Harry Potter (fictional character), 262
Hawking, Stephen, 126*n*
hawks, 177–78, 178*n*
haze, of space, 151–52
heat. *See also* cold; fire
  air and, 13–14, 31, *31*, 44–45, *45*
  in deep ocean, 159–60
  Earth, core of, producing, 227
  iron and, *16*, 16–18, *17*, *18*, 18*n*
  of Jupiter, 71–73
  radiation of, *12*, 12–13, *13*
  refrigerators and, 206–7, *208*, 209
  size and, 104–5, *105*
  of space, 93–94
  Sun, cooling of, 323–28
  of Sun, *105*, 105–6, 314–18, 314*n*, 316*n*, 321–23
  toasters, heating house with, 234–35, *235*
  of vacuum tube phone, *190*, 190–91
helicopters
  MRI scanners and landing, 171
  rotors of, 6–10, *7*, 8*n*
helium, 300–301, 301*n*
high seas, 127
Homer, 208
house, toasters heating, 234–35, *235*
house floor, as air hockey table, 230
houses of worship, 52–54
humans, 97
  descent from total, 172–75, *173*, *174*, *175*
  economic value of human lives, 121
  human lives, total number of, 24–25
  metabolism of, 208–9, *209*, 299
  nutritional value of, as food, 37–39

I

Icarus (mythological figure), 317
ice sheets, melting, 246*n*
identical ancestors point, 173–75
Illinoian glaciation, 282
impact tsunami, 247, 247*n*
insects, 90, *90*, 96–97. *See also* ants; bees
iPhone. *See* phone, made from vacuum tubes
iron, *16*, *17*
  in atmosphere, *18*, 18–20
  ultracold, 11, 14–15, *15*
  vaporizing, 17–21, 17*n*, *18*, 18*n*, *21*
IRS, 261
isotopes, stability of, 104, 107

J

Japan, *244*
  disappearance of, 244–48, 245*n*, *246*, *247*
  weight of, 245–46
jellyfish, 229
jet pack, 34, *35*
Johnson, Dwayne, 208–9, *209*
Jupiter, *1*, 1–3, *72*
  shooting through, 138
  shrinking, 70–73
Jurassic Park (fictional park), 102

K

Kalt, Brian C., 95
Keeler, Cindy, 237–39
Kelvin, 11–12, *12*
Kepler-1606, 24
Kevlar, 183
kissing, 135

L

L1 Lagrange point, 289–90
Lake Toba, eruption creating, 280–81, 307*n*
lamp bulbs, high-intensity, 202–3, 203*n*
landslide, 100
Large Hadron Collider (LHC), 184
large igneous provinces, 138
lasers, *43*, 43–46
  baking with, 228
  umbrella using, 192–94

lava, 62. *See also* volcanoes
extinction and, 138–39, 138*n*
lava lamp made from, 202–5, *203*
Mount Everest turning into, 138–39
laws
necessity of, 259–60
private letter ruling on, 261
reading, 255–59, 256*n*, 257*n*, 261
leaf blowers, 89
Lee, Harper, 48–49
lemon drops, rain of gumdrops and, 329–44
lenses, 250–53, 251*n*, *252*
LHC. *See* Large Hadron Collider
libraries, law, 259
Library of Alexandria, 47
light, *253*, 253*n*. *See also* sunlight
in atmosphere, *44*, 44–46, 44*n*
in deep ocean, 160–61
lenses concentrating, 250–53, 251*n*, *252*
moonlight, starting fire with, 249–51, 254
space distorting, 152
sugar emitting, 223–25
light-nanosecond, 314*n*
lightning, 223–25
lips, 135
liquid nitrogen, 14, 262
liquid oxygen, 13
liquid uranium, 30
Liu, Ting Ting, 26
Lowe, Derek, 285
LSD, 167
lungs, 19

M

magnetic fields, 168–71, 228
magnifying glass, 249–54
Mahowald, Natalie, 18–20
mailbox, punching, 185
man-made objects, average size of, 303
Mariana Trench, 140, 159–63
Mario (video game character), 33, *33*
Martin, John, 20
mass, of Earth, 59–65
Massachusetts, 256–57
match, striking, 141–42
McDonald's, 39, 39*n*
mega-skyscrapers. *See* buildings, size of
melting, 16–17, 31, 230, 246*n*, 317
mercury (element), 103, 204, 204*n*
Mercury (planet), 103
metabolism
dinosaur, 37–39, 38*n*
human, 208–9, *209*, 299
metals, expensive, 165
meteors, 99–101
dinosaurs, extinction of, and, 98, 102, 138*n*, 215
Earth, spin of, and, 214–15, *215*
Michelangelo, 157, 157*n*
microplastics, 129–30
Micro SD cards, 164, 167
microwave, 91, 298
Milky Way, 3–4, 125–26, *126*
Earth relative to, 124, *124*
flying across, 231–32
stars in, 74–75, 77
visibility of, 152, *152*
mines, 159–60
mirage, 45, *45*
mirrors, 51, 242–43, 250
MIS-8 glacial period, 282
Mississippi River, 281
molten glass, 203, 203*n*
momentum, of bullet, 57
Moon, 106, 110, 216, 289*n*, *294*, *295*
destroying, 228
to Earth, pole from, 287–97, *289*, *293*
of electrons, 236–40, 239*n*
moonlight, starting fire with, 249–51, 254
moon jellies, 229
Morgan le Fay (fictional character), 46, *46*
Moses Mabhida Stadium (Durban, South Africa), 81
Mount Cayambe, 293*n*
Mount Chimborazo, 293*n*
Mount Everest, 148, 293*n*
as lava, 138–39
snowball rolling down, 268–71
MRI medical scanners, 168–71
Muller, Derek, 220*n*
murder, 94–95
mushroom cloud, 72

N

naked singularity, 239
Nanjing University of Aeronautics and Astronautics, 26
NASA, 23*n*
National Weather Service, 306–7
Neptune, 103–7, *106*
neptunium, 103–7, *106*, 106*n*
New Horizons (spacecraft), 23*n*
New York City
time running backward in, 277, 283
T. rex in, 37–39
Niagara Falls, 272–76, 274*n*
nictitating membrane, 243*n*
nitrogen, 14, 31, 262
no hair theorem, 5

O

Obama, Barack, 231–32
ocean
currents in, 247, *247*
iron in, 20, *21*
tube in, 159–63
The Odyssey (Homer), 208
oil, 121*n*, 131–34
oil, engine, 23, *23*, 23*n*
Old Faithful, 40–42
Olson, Ted, 122–23
101 Dalmatians (film), 309–10, 313
open-bottom container, water and, 143–48
oxidizer, 141–42, 141*n*
oxygen, 13–14, 31, *31*, 141–42, 141*n*
ozone, 302–3

P

paint, 66–69
Parker, G. A., 27*n*
particle accelerators, 184, 237–38
pay-to-play mobile game, 164–65
peanut butter, 111
pendulum, 78–80
Pennycuick, C. J., 27*n*
Permian extinction, 139
phone, made from vacuum tubes, 186–91, *190*
photosynthesis, 132*n*, 134
pigeons
  flight of, 26–29, 27*n*, 231, *231*
  in space, 231, *231*
  weight carried by, 26–29, *29*
plastic, 131, 134
plate tectonics, 227
platinum, 165, 167
plica semilunaris, 243, 243*n*
Pluto, 103–4
plutonium, 103–5, 165
poké balls, 260
pole climbing, 288–90, 288*n*
Portsmouth Gaseous Diffusion Plant, 106
potassium chlorate, 141*n*
potatoes, melting and, 230
private letter ruling, 261
property values, 68, 155, 155*n*
protons, 142, 236–37, 239*n*
puppies, 308–13, 309*n*

Q

Q1 skyscraper (Australia), 26, 28–29

R

rabies, 226
radiation, of heat, *12*, 12–13, *13*
radiation, UV, 319–21
rain
  lasers stopping, 192–94
  of lemon drops and gumdrops, 329–44
red dwarf, 75–76
red giant, 76
refrigerators, 206–7, *208*, 209
regulations, 256
Reissner-Nordström equations, 239
relativity, 237–39, 237*n*
religious observance, survey of, 52–53
resolution, 150, *150*
road, into space, 137
rocket, 138
Rohde, Douglas L. T., 173–74
Rome, building of, 154–58, 155*n*, *156*
roofs, 331
room temperature, stars at, 24, 323–24, 324*n*
rotation, of Earth, 108–10, 108*n*
rotors, helicopter. *See* helicopters
rubber tires, 128–30

S

sails, 140–41
saliva, 263–67
sand, 74–77, 99–100
saponification, 285–86
sapphire, 203
satellites, 109
sea level, 245–48, *246*, 246*n*
Shannon, Claude, 186
sharks, 145–46
ship, sails of, 140–41
shoebox, expensive way of filling, 164–67
Sistine Chapel, 157, 157*n*
Sisyphus (mythological figure), 206, *208*, 208–9
size, heat and, 104–5, *105*
skydiving, 34, *34*
skyscrapers. *See* buildings, size of; Q1 skyscraper
smartphone. *See* phone, made from vacuum tubes
smell, of stars, 301–2
snakes, 33, *33*
snow
  Earth covered with, 305–7
  snowballs, 268–71, *270*, *271*, 307
soil liquefaction, 100
solar panels, 60, *60*
Solar System, *1*, 110
  soup filling, 1–5
  survival in, 232
soup, 1–5
space
  anvil dropped from, 233
  haze of, 151–52
  heat in, 93–94
  pigeons in, 231, *231*
  road into, 137
  visible objects in, *152*, 152–53
space elevator, 119, 137
space heaters, 234–35, *235*
speed
  driving, 180–81, *181*
  Earth, rotation of, speeding up, 108–10
  of helicopter rotors, 6–9
  of Moon, 293–95, *294*, *295*
  writing, 48–50
SPF, 318–21, 321*n*
spiders, 217–19, 218*n*, 219, 219*n*
spin, of Earth, 214–16, *215*, 287, 293, 293*n*
stamps, 122, 122*n*
Stapp, John Paul, 181
stars, 126–27
  collapsing, 152–53
  in Milky Way, 74–75, 77
  room-temperature, 24, 323–25, 324*n*
  sizes of, 75–76
  smell of, 301–2
  taste of, 303
  white dwarf, 323–28, *324*, 324*n*, *327*
statutes, 256
Stefan-Boltzmann law, 190, *190*, 317
stomach, ammonia in, 284–86
Stommel, Henry, 163
straw, Niagara Falls in, 272–76
string theory, 237–38
sugar, 223–25, 332
Sun, 160–61. *See also* solar panels; Solar System
  cooling of, 323–28
  Earth consumed by, 323*n*
  gravitational pull of, 217–19, 323
  heat of, *105*, 105–6, 314–18, 314*n*, 316*n*, 321–23

Sun (*cont.*)
insects flying to, 90, *90*
interior of, 316–17
surface of, 314–16, 314*n*, 318–22
taller people and, 199, 201
time running backward and, 277–78
two sunrises, sunsets, seeing, 200, *200*
UV radiation from, 319–21.
as white dwarf, 323–28, *324*, *327*
sunlight, 51, 315–16
in atmosphere, *44*, 44–46, 44*n*
energy from, *60*, 60–61
in fossil fuels, 132, 132*n*
lenses concentrating, 250–53, 251*n*, *252*
sunscreen, 318–22, 321*n*
Super Mario Bros. (video game), 33, *33*
Supreme Court, 257, 257*n*
SUV, leaf blowers moving, 89
swim bladders, 145, 147
swimming pools, 263–67, 264*n*
swing set, 78–82
sword, of air, 31, *31*

T

taller people, 199–201
Taskmaster (television show), 300*n*
taste, of stars, 303
teeth, 185
telescopes, 150–51, 252*n*
Tellado, Corín, 48–49, *49*
tensile strength, 270, *270*
tephra, 204, 204*n*
terminal velocity, 233
thawing, 14–15
thermodynamics, 250–51, 252*n*
tidal forces, 3
time, walking backward in, 277–83
tire rubber, 128–30
Titan, 141–42
toasters, heating house with, 234–35, *235*
tonsils, 51
trampoline, 92
transistors, 186, *186*, *187*
treaties, 255–57
Treskilling Yellow postage stamp, 122
T. rex. *See Tyrannosaurus rex*
triboluminescence, 223–25
trillion-dollar platinum coin, 167
tsunamis, *101*, 101–2, 247–48, 247*n*, 266
tube, in ocean, 159–63
Tyrannosaurus rex (T. rex), 37–39, 37*n*, 38*n*

U

ultracold materials, 11, 14–15, *15*
umbrella, laser, 192–94
undecillion, 120, 123
UNIVAC, *187*, 187–89
universe
black holes and, 239–40, 239*n*
edge of observable, 22–25
stars in visible, 74
uranium, 30, 103–5, 107
Uranus, 103–5, 107
UV radiation, 319–21

V

vacuum, 93, 230
vacuum cleaner, 89, 185
vacuum tubes, *186*
in computers, 186–87, *187*
phone made from, 186–91, *190*
vapor, water, 333–34
vaporizing. *See also* evaporation
iron, 17–21, 17*n*, *18*, 18*n*, *21*
water, 193, 193*n*
Venus, 232–33, 334
visibility distance, 150, *150*
visible objects, in space, *152*, 152–53
Vladivostok, Russia, 247, 247*n*
volcanoes, *61*, 61–62, 204, 280–81, 307*n*

W

walking backward, in time, 277–83
water, 192, 194. *See also* rain; snow
in body, 31, *32*, 134
in clouds, 195–96
container of, throwing, 299
density of, 71, 197, 301, 301*n*, 332, 332*n*
dinosaurs in, 134
Earth without, 34, *34*
flow rate of, 273–75, 273*n*
open-bottom container and, 143–48
vaporizing, 193, 193*n*
weight of, 301, 301*n*
water vapor, 333–34
weight
body, Earth and, 59–65, 108, 108*n*
buildings, size of, and, 111–12
of Japan, 245–46
pigeons carrying, 26–29, *29*
of water, 301, 301*n*
welder, 35
Wentworth, Chester K., 74–75
whipped cream, 197, 197*n*
white dwarf, 324*n*
gravity on, 326–27, *327*
Sun as, 323–28, *324*, *327*
Whittlesey, Lee H., 40–41
Wilson, Tracy V., 197*n*
wind, 112–13, 140–41
Wint-O-Green Life Savers, 223, 223*n*
The Wizard of Oz (film), 167
world population, assembling, 155, 155*n*
worship, houses of, 52–54
writing, speed of, 48–50

Y

Yellowstone National Park, *40*, 40–42, *41*, 95

Z

Zittrain, Jonathan, 259, 261